T0182014

Understanding and Using Rough Set Based Feature Selection: Concepts, Techniques and Applications

Muhammad Summair Raza ·
Usman Qamar

Understanding and Using Rough Set Based Feature Selection: Concepts, Techniques and Applications

Second Edition

 Springer

Muhammad Summair Raza
Department of Computer and Software
Engineering, College of Electrical
and Mechanical Engineering
National University of Sciences
and Technology (NUST)
Islamabad, Pakistan

Usman Qamar
Department of Computer and Software
Engineering, College of Electrical
and Mechanical Engineering
National University of Sciences
and Technology (NUST)
Islamabad, Pakistan

ISBN 978-981-32-9168-3   ISBN 978-981-32-9166-9   (eBook)
https://doi.org/10.1007/978-981-32-9166-9

This Springer imprint is published by the registered company Springer Nature Singapore Pte Ltd.
The registered company address is: 152 Beach Road, #21-01/04 Gateway East, Singapore 189721, Singapore

This book is dedicated to the memory of Prof. Zdzisław Pawlak (1926–2006), the father of Rough Set Theory. He was the founder of the Polish school of Artificial Intelligence and one of the pioneers in Computer Engineering and Computer Science.

Preface

Rough set theory, proposed in 1982 by Zdzislaw Pawlak, is in constant development. It is concerned with the classification and analysis of imprecise or uncertain information and knowledge. It has become a prominent tool for data analysis. This book provides a comprehensive introduction to rough set-based feature selection. It enables the reader to systematically study all topics of rough set theory (RST) including preliminaries, advance concepts, and feature selection using RST. This book is supplemented with RST-based API library that can be used to implement several RST concepts and RST-based feature selection algorithms.

The book is intended to provide an important reference material for students, researchers and developers working in the areas of feature selection, knowledge discovery and reasoning with uncertainty, especially for those who are working in RST and granular computing. The primary audience of this book is the research community using rough set theory (RST) to perform feature selection (FS) on large-scale datasets in various domains. However, any community interested in feature selection such as medical, banking, finance can also benefit from the book.

The second edition of the book now also covers dominance-based rough set approach and fuzzy rough sets. Dominance-based rough set approach (DRSA) is an extension to the conventional rough set approach which supports the preference order using dominance principle. Fuzzy rough sets are fuzzy generalization of rough sets. API library of dominance-based rough set approach is also provided with the second edition of the book.

Islamabad, Pakistan

Muhammad Summair Raza
Usman Qamar

Contents

About the Authors

Muhammad Summair Raza has Ph.D. specialization in software engineering from the National University of Sciences and Technology (NUST), Pakistan. He completed his MS from International Islamic University, Pakistan, in 2009. He is also associated with the Virtual University of Pakistan as assistant professor. He has published various papers in international-level journals and conferences with a focus on rough set theory. His research interests include feature selection, rough set theory, trend analysis, software architecture, software design and non-functional requirements.

Dr. Usman Qamar is currently a tenured associate professor at National University of Sciences and Technology (NUST) having over 15 years of experience in data engineering and decision sciences both in academia and industry having spent nearly 10 years in the UK. He has a Masters in Computer Systems Design from University of Manchester Institute of Science and Technology (UMIST), UK. His M.Phil. in Computer Systems was a joint degree between UMIST and University of Manchester which focused on feature selection in big data. In 2008/09, he was awarded Ph.D. from University of Manchester, UK. His Ph.D. specialization is in Data Engineering, Knowledge Discovery and Decision Science. His Post Ph.D. work at University of Manchester, involved various research projects including hybrid mechanisms for statistical disclosure (feature selection merged with outlier analysis) for Office of National Statistics (ONS), London, UK, churn prediction for Vodafone UK and customer profile analysis for shopping with the University of Ghent, Belgium. He has also done a post graduation in Medical and Health Research, from University of Oxford, UK, where he worked on evidence-based health care, thematic qualitative data analysis and healthcare innovation and technology. He is director of Knowledge and Data Science Research Centre, a Centre of Excellence at NUST, Pakistan and principal investigator of Digital Pakistan Lab, which is part of National Centre for Big Data and Cloud Computing. He has authored over 150 peer reviewed publications which includes 2 books published by Springer & Co. He has successfully supervised 5 Ph.D. students and over 70 master students. Dr. Usman has been able to acquire nearly PKR 100 million in research grants. He has received multiple research awards, including Best Researcher of

Pakistan 2015/16 by Higher Education Commission (HEC), Pakistan as well as gold in Research and Development category by Pakistan Software Houses Association (P@SHA) ICT Awards 2013 and 2017 and Silver award in APICTA (Asia Pacific ICT Alliance Awards) 2013 in category of R&D hosted by Hong Kong. He is also recipient of the prestigious Charles Wallace Fellowship 2016/17 as well as British Council Fellowship 2018, visiting research fellow at Centre of Decision Research, University of Leeds, UK and visiting senior lecturer at Manchester Metropolitan University, UK. Finally, he has the honour of being the finalist of the British Council's Professional Achievement Award 2016/17.

Chapter 1
Introduction to Feature Selection

This is an era of information. However, the data is only valuable if it is efficiently processed and useful information is derived out of it. It is now common to find applications that require data with thousands of attributes. Problem with processing such datasets is that they require a huge amount of resources. To overcome this issue, the research community has come up with an effective tool called feature selection. Feature selection lets us select only relevant data that we can use on behalf of the entire dataset. In this chapter, we will discuss necessary preliminaries of feature selection.

1.1 Feature

A feature is a characteristic or an attribute of an object where an object is an entity having physical existence. A feature is an individual measurable property of a phenomenon being observed [1]. For example, features of a person can be 'Height', 'Weight', 'Hair Color' etc. A feature or combination of features helps us perceive about a particular aspect of that object, e.g. the feature 'Height' of a person helps us visualize physical height of the person. Similarly, the feature 'Maximum Load' gives us the information about the maximum upper bound of the force that a glass can bear with. For a particular concept or model, the quality of features is important. The better the quality of feature, the better is the resulting model [2].

In a dataset, data is represented in the form of matrix of 'm' rows and 'n' columns. Each row represents a record or object, whereas each column represents a feature. So, in a dataset, each object is represented in the form of collection of features. For example, the dataset in Table 1.1 represents two features of five people.

© Springer Nature Singapore Pte Ltd. 2019
M. S. Raza and U. Qamar, *Understanding and Using Rough Set Based Feature Selection: Concepts, Techniques and Applications*,
https://doi.org/10.1007/978-981-32-9166-9_1

Table 1.1 Sample dataset

Name	Height	Eyes colour
John	5.7	Blue
Bill	6.1	Black
Tim	4.8	Black
David	4.5	Blue
Tom	5.1	Green

It should be noted that all the data may not be in the form of such an arranged structure of rows and columns, e.g. DNA and protein sequences, in which case you may need to extract the features using different feature extraction techniques.

Based on the nature (domain) of data and type of the values a feature may take, features can be classified into two types:

(a) Numerical,
(b) Categorical.

1.1.1   Numerical

Numerical features, as a simple definition are those that take numbers as their values. For example, in case of height, 5 feet and 7 inches is a numerical value, (however, if we mention height as tall, medium and short then the height will not be a numerical feature). Based on the range of values, numerical features can be discrete or continuous. Discrete values can be finite or infinite. Finite features are the ones that take a limited set of values, e.g. 'Total Matches played', 'Total Scores Made'. Infinite features are the ones for which the value set is infinite, e.g. total number of coin flips that result heads. Continuous values, on the other hand, belong to the domain of real numbers. For example, height of a person, weight in grams, etc.

Numerical features can be further classified into two types:

Interval-scaled: Interval-scaled features are the ones where the difference between the two values is meaning. For example, the difference between 90° and 70° is same as the difference between 70° and 50° of temperature.

Ratio-scaled: Ratio-scaled features have all the properties of interval-scaled features plus defined ratios for data analysis. For example, for attribute Age, we can say that someone who is 20 years old is twice as old as someone who is 10 years old.

1.1.2   Categorical Attributes

In categorical attributes, symbols (words) are used to represent domain values. For example, gender can be represented by two symbols 'M' and 'F' or 'Male' or 'Female'. Similarly, qualification can be represented by 'Ph.D.', 'BS' and 'MS', etc. Categorical attributes can be further classified into two types (Fig. 1.1).

Nominal: Nominal features are the ones where order does not make sense. For example, the feature 'Gender' is a nominal feature because the domain values 'M' and 'F' does not involve any order. So, the comparison here only involves comparing domain values. Furthermore, only, equal operator makes sense, i.e. we cannot compare values for less-than or greater-than operators.

Ordinal: In ordinal features, both equality and inequality can be involved, i.e. we can use 'Equal-to', 'Less-than' and 'Greater-than' operators. For example, 'Qualification' feature is an example of ordinal type as we can involve all of the above operators in comparison.

Figure 1.2 shows the categorization of types of features.

1.2   Feature Selection

The amount of data to be processed and analyzed to get conclusions and to make decisions has significantly increased these days. Data creation is occurring at a record rate [3]. Data size are generally growing from day to day [4] and the rate at

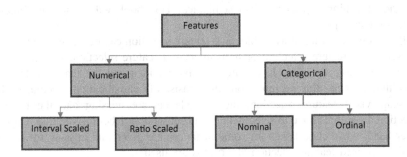

Fig. 1.1   Categorization of feature types

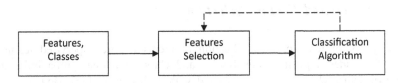

Fig. 1.2   Generic process of supervised feature selection

which new data are being generated is staggering [5]. This increase is seen in all areas of human activity right from the data generated on daily basis such as telephone calls, bank transactions, business tractions to more technical and complex data including astronomical data, genome data, molecular datasets, medical records, etc. It is amount of information in the world doubles every twenty (20) months [6]. These datasets may contain a lot of information useful but still undiscovered.

This increase in data is twofold, i.e. both in number of samples/instances and number of features that are recorded and calculated. As a result, many real-world applications need to process datasets with hundreds and thousands of attributes. This significant increase in the number of dimensions in datasets leads to a phenomenon called curse of dimensionality. The curse of dimensionality is the problem caused by the exponential increase in volume associated with adding extra dimensions to a (mathematical) space [7].

Feature Selection is one of the solutions to the dilemma of curse of dimensionality. It is the process of selecting a subset of features from the dataset that provides most of the useful information [6]. The selected set of features can then be used on behalf of the entire dataset. So, a good feature selection algorithm should opt to select the features that tend to provide complete or most of the information as present in the entire dataset and ignore the irrelevant and misleading features. Dimensionality reduction techniques can be categorized into two methods 'Feature Selection' and 'Feature Extraction'. Feature extraction techniques [8–20] project original feature space to a new feature space with lesser number of dimensions. The new feature space is normally constructed by combining the original feature some way. The problem with these approaches is that the underlying semantics of data is lost. Feature selection techniques [21–34], on the other hand, tend to select features from the original features to represent the underlying concept. This lets feature selection techniques preserve data semantics. This book will focus on Feature selection techniques.

Based on the nature of available data, feature selection can be categorized either as supervised feature selection or non-supervised feature selection. In supervised feature selection, the class labels are already provided and the feature selection algorithm selects the features on the basis of classification accuracy. In non-supervised feature selection, the class labels are missing and the feature selection algorithms have to select feature subset without label information. On the other hand, when class labels for some instances are given and missing for some, semi-supervised feature selection algorithms are used.

1.2.1 Supervised Feature Selection

The majority of real-world classification problems require supervised learning where the underlying class probabilities and class-conditional probabilities are unknown, and each instance is associated with a class label [35]. However, too often we come across the data having hundred and thousands of features. Presence of noisy, redundant and irrelevant features is another problem we have to commonly face in the classification process. A relevant feature is neither irrelevant nor redundant to the target concept; an irrelevant feature is not directly associate with the target concept but affect the learning process, and a redundant feature does not add anything new to the target concept [35]. Normally, it is difficult to filter out the important features especially when the data is huge. All this affects the performance and accuracy of classifier. So, we have to preprocess data before feeding to the classification algorithm. Supervised Feature Selection, at this point plays its role to select minimum and relevant features. Figure 1.2 below shows the general process of supervised feature selection.

As discussed earlier, in supervised feature selection, the class labels are already given along with features where each instance belongs to an already specified class label and feature selection algorithm selects features from the original features based on some criteria. The selected features are then provided as input to the classification algorithm. The selection of features can or cannot be independent of the classification algorithm based on either it is filter-based or wrapper-based approach. This is realized by dotted arrow in the above diagram from 'Classification algorithm' to 'Feature Selection' process. Because the selected features are fed to classification algorithm, quality of selected features affects the performance and accuracy of the classification algorithm. The quality feature subset is the one that produces the same classification structure that is otherwise obtained by using the entire feature set.

Table 1.2 below shows a simple supervised datasets.

Here, {A, B, C, D} are normal features and 'Z' is class label. We can perform feature selection by using many criteria, e.g. information gain, entropy, dependency. If we use Rough Set-Based dependency measure, then dthe ependency of 'Z' on {A, B, C, D} is '100%', i.e. the feature set {A, B, C, D} uniquely determines the value of 'Z'. On the other hand dependency of 'Z' on {A, B, C} is also '100%', it means we can skip the feature 'D' and use {A, B, C} only for further processing. A dependency simply determines how uniquely the value of an attribute 'C' determines the value of an attribute 'D'. Calculation of Rough Set Based Dependency measure is discussed with details in Chap. 33.

Table 1.2 A sample dataset in supervised learning

U	A	B	C	D	Z
X1	L	3	M	H	1
X2	M	1	H	M	1
X3	M	1	M	M	1
X4	H	3	M	M	2
X5	M	2	M	H	2
X6	L	2	H	L	2
X7	L	3	L	H	3
X8	L	3	L	L	3
X9	M	3	L	M	3
X10	L	2	H	H	2

1.2.2   Unsupervised Feature Selection

It is not necessary that classification information is given all the time. In unsupervised learning, only the features are provided without any class label. The learning algorithm has to use only the available information. So, a simple strategy may be to form the clusters (a cluster is a group of similar objects, similar to a classification structure except that class labels are not provided and hence only original features are used to form clusters, whereas in forming classification structures, class labels are given and used). Now again the problem of noisy, irrelevant and redundant data prohibits the use of all of the features to feed to learning algorithm, furthermore removing such features is again a cumbersome task for which manual filtration may not be possible. Again, we have to take help from the feature selection process. So, a common criteria to select those features that give the same clustering structure as given by the entire set of features.

Consider the following example using hypothetical data. Table 1.3 below shows the unsupervised dataset.

Dataset contains objects {A, B, C, D} where each object is characterized by features F = {X, Y, Z}

Using K-Means clustering, the objects {A, D} belong to cluster C_1 and objects {B, C} belong to cluster C_2. Now if computed, the feature subsets {X, Y}, {Y, Z} and {X, Z} can produce the same clustering structure. So any of them can be used as the selected feature subset. It should be noted that we may have different feature subsets that fulfil the same criteria, so any of them can be selected but efforts are made to find the optimal one, the one having minimum number of features.

Following is a description of how K-means clustering theorem works and how we calculated clusters.

K-Means Clustering Theorem:
K-means clustering theorem is used to calculated clusters in unsupervised datasets. The steps of the theorem are given below:
Do {
Step-1: Calculate centroid
Step-2: Calculate distance of each object from centroids
Step-3: Group objects based on minimum distance from centroids
} until no objects moves from one group to other.

Table 1.3 A sample dataset in unsupervised learning

	X	Y	Z
A	1	2	1
B	2	3	1
C	3	2	3
D	1	1	2

First of all, we calculate the centroids, a centroid is centre point of a cluster. For the first iteration, we may assume any number of points as centroids. Then we calculate the distance of each point from centroid, here distance is calculated using Euclidean distance measure. Once the distance of each object from each centroid is calculated, the objects are arranged in such a way that each object falls in the cluster whose centroid is close to that point. The iteration is repeated then by calculating new centroids of every cluster using all the points that fall in that cluster after an iteration is completed. The process continues until no point changes its cluster.

Now consider the data points given in above datasets:

Step-1: We first suppose point A and B as two centroids C_1 and C_2.

Step-2: We calculate Euclidean distance of each point from these centroids using equation of Euclidean distance as follows:

$$d(x\,y) = \sqrt{(x_1 - x_2)^2 + (y_1 - y_2)^2}$$

Note: x_1, y_1 are coordinates of point X and x_2, y_2 are coordinates of point Y. In our dataset, the Euclidean distance of each point will be as follows:

$$D^1 = \begin{bmatrix} 0 & 1.4 & 2.8 & 1.4 \\ 1.4 & 0 & 2.4 & 2.4 \end{bmatrix} C_1 = \{1, 2, 1\} \ and \ C_2 = \{2, 3, 1\}$$

Using 'Round' function:

$$D^1 = \begin{bmatrix} 0 & 1 & 3 & 1 \\ 1 & 0 & 2 & 2 \end{bmatrix} C_1 = \{1, 2, 1\} \ and \ C_2 = \{2, 3, 1\}$$

In the above distance matrix, the value at row = 1 and column = 1 represents the distance of point A from first centroid (here, point A itself is the first centroid, so it is distance of point A with itself which is zero), the value at row = 1 and column = 2 represents the Euclidean distance between points B and first centroid, similarly, the value at row = 2 and column = 1 shows Euclidean distance between point A and second centroid and so on.

Step-3: If we look at column three, it shows that point 'C' is close to the second centroid as compared to its distance from the first centroid. On the basis of above distance matrix, the following groups are formed:

$$G^1 = \begin{bmatrix} 1 & 0 & 0 & 1 \\ 0 & 1 & 1 & 0 \end{bmatrix} C_1 = \{1, 2, 1\} \ and \ C_2 = \{2, 3, 1\}$$

where the value '1' shows that point falls in that group, so from above group matrix, it is clear that point A and D are in one group (cluster), whereas points B and C are in other group (cluster).

Now, the second iteration will be started. As we know that in the first cluster, there are two points 'A' and 'D', so we will calculate their centroid as the first step. So, $C1 = \frac{1+1}{2}, \frac{2+1}{2}, \frac{1+2}{2} = 1, 2, 2$. Similarly, $C_2 = \frac{2+3}{2}, \frac{3+2}{2}, \frac{1+3}{2} = 2, 2, 2$. On the basis of these new centroids, we calculate the distance matrix which is as follows:

$$D^2 = \begin{bmatrix} 1 & 2 & 2 & 1 \\ 1 & 1 & 1 & 1 \end{bmatrix} C_1 = \{1, 2, 2\} \text{ and } C_2 = \{2, 2, 2\}$$

$$G^2 = \begin{bmatrix} 1 & 0 & 0 & 1 \\ 0 & 1 & 1 & 0 \end{bmatrix} C_1 = \{1, 2, 2\} \text{ and } C_2 = \{2, 2, 2\}$$

Since $G^1 = G^2$, so we stop here. Note that the first column of D^2 matrix shows that point A is at equal distance from both clusters C_1 and C_2, so it can be placed in any of the clusters.

The above-mentioned clusters were formed considering all features {X, Y, Z}. Now if you perform the same steps using feature subsets {X, Y} or {Y, Z} or {X, Z}, same clustering structure is obtained, which means that any of the above feature subsets can be used by the learning algorithm.

1.3   Feature Selection Methods

Based on the relationship between selected features and learning algorithm or evaluation of selected feature subset, feature selection methods can be divided into three categories:

- Filter Methods,
- Wrapper Methods,
- Embedded Methods.

1.3.1   Filter Methods

Filter method is the most straightforward strategy for feature selection. In this method, selection of features remains independent of learning algorithm, i.e. no feedback from the classification algorithm is used. Features are evaluated using some specific criteria using the intrinsic properties of the features. So, the classification algorithm has no control over the quality of selected features and thus poor quality may affect the performance and accuracy of the subsequent algorithm. They can be further classified into two types [36]:

1. Attribute evaluation methods,
2. Subset evaluation methods.

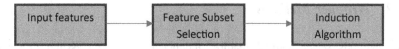

Fig. 1.3 Generic process of filter-based feature selection process

```
Input:
S - data sample with features X,|X| = n
J - evaluation measure to be maximized
GS – successor generation operator
Output:
Solution – (weighted) feature subset
L := Start_Point(X);
Solution := { best of L according to J };
repeat
L := Search_Strategy (L,GS(J),X);
X' := {best of L according to J };
if J(X')≥J(Solution) or (J(X')=J(Solution) and |X'| < |Solution|) then Solution :=X';
until Stop(J,L).
```

Fig. 1.4 Generic filter-based algorithm [37]

Attribute evaluation methods evaluate each individual feature according to the selected criteria and rank each feature after which a specific number of features are selected as the output. The subset evaluation methods, on the other hand, evaluate the complete subset on the basis of specified criteria. Figure 1.3 shows the generic diagram of filter-based approach [38].

You can note the unidirectional link between feature selection process and induction algorithm. Figure 1.4 shows the pseudocode of a generic feature selection algorithm.

We have dataset having feature set X with n number of features. 'J' is the measure that is maximized, it is basically the evaluation criteria on the basis of which features are selected, e.g. dependency, information gain, relevancy. GS represents the successor generation operators which is used to generate next feature subset from the current one. This operator may add or delete features from the current subset based on 'J' measure. 'Solution' is the subset that will contain the optimized feature subset output. Initially we assign 'Solution' a starting feature subset. The starting subset may be empty (in case of forward feature selection strategy), may contain entire feature subset (in case of backward feature selection strategy) or random features (in case of random algorithms).

We get next 'L' features from the initially assigned feature subset, then optimize 'L' into 'X'. Solution is then assigned 'X' if 'X' is more appropriate than previous one. This step ensures that 'Solution' always contains optimized feature subset and that we gradually refine the feature subset. The process continues until we meet the stopping criteria.

1.3.2   Wrapper Methods

Filter methods select optimal features independent of the classifier, however, optimal and quality feature subsets are dependent on the heuristics and biases of the classification algorithm, so should be aligned with and selected on the basis of underlying classification algorithm. This is the underlying concept of wrapper approaches. So, in contrast with the filter methods, wrapper approaches do not select features independent of the classification algorithm. The feedback of the classification algorithm is used to measure the quality of selected features and thus results in high quality and performance of classifier. Figure 1.5 shows the generic model of wrapper approach [38].

It can be seen that there is a two-way link between 'Feature Subset Search', 'Evaluation' and 'Induction' processes, i.e. features are evaluated and thus searched again on the basis of the feedback from induction algorithm. From the figure, it is clear that wrapper approaches consist of three steps:

Step-1   Search feature subset,
Step-2   Evaluate features on the basis of induction algorithm,
Step-3   Continue process until we get optimized feature subset.

Clearly, induction algorithm works as a black box where the selected feature subset is sent and the acknowledgement is received in the form of some quality measure, e.g. error rate.

1.3.3   Embedded Methods

Embedded methods tend to overcome the drawbacks of both filter and wrapper approaches. Filter methods evaluate features independent of classification algorithm, while wrapper-based approaches, using feedback from classifier, are computationally expensive as classifier is run many times to select optimal feature subset. Embedded models construct feature subsets as part of the classifier, so they

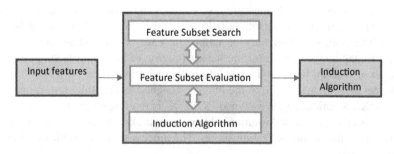

Fig. 1.5 Generic process of wrapper based feature selection process

observe advantages of both wrapper approaches (feature subset evaluation is not independent of classifier) and filter method (efficient than wrapper approaches, furthermore selected features are evaluated using independent measures as well). A typical embedded method works as follows:

Step-1   Initialize feature subset (either empty or containing all the features).
Step-2   Evaluate the subset using independent measure.
Step-2.1 If it is fulfils criteria more than current subset, this becomes current subset.
Step-3   Evaluate the subset against evaluation criteria specified by classifier.
Step-3.1 if it is fulfils criteria more than current subset, this becomes current subset.
Step-4   Repeat Step-2–Step-3 until criteria is met.

There are three types of embedded methods [39]. The first is pruning methods that initially train the model using entire set of features and then remove features gradually, then there are models that provide build-in mechanism to perform feature selection and finally, there are regularization models that minimize fitting errors and simultaneously remove the features by forcing the coefficients to be small or zero.

Table 1.4 shows the advantages and disadvantages of each approach.

1.4   Objective of Feature Selection

The main objective of feature selection is to select a subset of features from entire dataset that could provide the same information otherwise provided by entire feature set. However, different researchers describe feature selection from different perspectives. Some of these are:

1. Faster and more cost-effective models: Feature selection tends to provide the minimum number of features to subsequent processes, so that these processes don't need to process the entire set of features, e.g. consider the classification with 100 attributes versus classification with 10 attributes. The reduced number of features means the minimum execution time for the model. For example, consider the following dataset given in Table 1.5.

   Now if we perform classification on the basis of attributes Temperature, Flue and Cough, we know that a patient having any of the symptoms S1, S2, S3, S4 and S6 will be classified as Sick. On the other hand, the patient with symptoms S5 or S7 will not be sick. So to classify a patient as sick or not, we have to consider all the three attributes. However, note that if we take only the attributes 'Temperature' and 'Flue', we can also correctly classify the patient as sick or not, consequently to decide about the patient, the selected features 'Temperature' and 'Flue' give us the same information (accurately classify the records) as otherwise obtained by the entire set of attributes. Figure 1.6 shows

Table 1.4 Comparison of filter, wrapper and embedded approaches

Feature selection method	Advantages	Disadvantages
Filter	Simple approach Computationally less expensive	No interaction with classifier, so the quality of selected features may affect the quality of classifier
Wrapper	Considers feature dependencies Feature selection involves feedback from classifier thus resulting in high-quality features	Computationally expensive than filter approach High overfitting probability
Embedded	Combines the advantages of both filter and wrapper approach	Specific to learning machine

Table 1.5 Symptoms table

Symptoms	Temperature	Flue	Cough	Sick
S1	H	Y	Y	Y
S2	H	Y	N	Y
S3	H	N	Y	Y
S4	N	Y	Y	Y
S5	N	N	N	N
S6	N	Y	N	Y
S7	N	N	Y	N

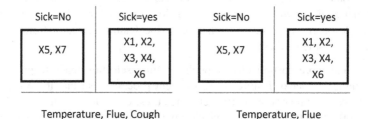

Fig. 1.6 Classification using entire feature set versus selected features

the classification obtained in both cases, i.e. using all features and using selected features.

Note that this was a very simple example but in the real world, we come across the applications that require to process hundreds and thousands of the attributes before performing further tasks, feature selection is a mandatory preprocessor for all such applications.

2. Avoid overfitting and improve performance: Selecting the best features that provide most of the information and by removing noisy, redundant and irrelevant features, the accuracy and effectiveness of model can be enhanced. It

reduces the number of dimensions and thus enhances the performance of the subsequent model. It helps reduce noisy, redundant and irrelevant features thus improving data quality and enhancing accuracy of the model.

If we consider the dataset given in Table 1.5, it can be clearly seen that using two features, i.e. 'Temperature' and 'flue' give us the same result as given by entire feature set, so ultimately the performance of the system will be enhanced. Furthermore, note that including or excluding the feature 'Cough' does not affect classification accuracy and hence the final decision, so, here it will be considered as redundant and thus can safely be removed.

3. Deeply understand the process that generated data: Feature Selection also provides an opportunity to understand the relationships between attributes to better understand the underlying process. And it helps to understand the relationship between the features and about the process that generated data. For example, in Table 1.5, it can be clearly seen that the decision class 'Sick' fully depends on the feature subset 'Temperature, Flue'. So, to accurately predict about a patient, we must have the accurate values of these two features.

It should be noted that no other combination of the features (except the entire feature set, i.e. including all three features) gives us the exact information. If we consider the combination 'Temperature, cough', we cannot decide about a patient having Temperature = N and Cough = Y (S4 and S7 lead to different decisions for same value of Temperature = N and Cough = Y). Similarly, if we combine 'Flue, Cough', we cannot decide about a patient having Flue = N and Cough = Y (S3 and S7 lead to different decisions for the same value of Flue = N and Cough = Y). However, for all any combination of 'Temperature, Flue', we can precisely reach the decision. Hence, there is a deep relationship between the decision 'Sick' and the features 'Temperature, Flue'.

It should also be noted that if an algorithm trained using the above dataset gets the input like Temperature = N, Flue = Y and Sick = No, it should immediately conclude that the provided input contains erroneous data, which means that there was some mistake in mechanism collecting the data.

1.5   Feature Selection Criteria

The core of feature selection is the selection of good feature subset, but what actually makes a feature subset good? That is, what may be the criteria to select features from the entire dataset. Various criteria have been defined in literature for this purpose. We will discuss a few here.

1.5.1 Information Gain

Information gain [40] can be defined in terms of 'Uncertainty'. The greater the uncertainty, the lesser is the information gain. If IG(X) represents information gain from feature X, then feature X will be more good (hence preferred) if IG(X) > IG (Y). If 'U' represents uncertainty function, P(Ci) represents class Ci probability before considering feature 'X' and P(Ci|X) represents posterior probability of class Ci considering the feature 'X', the information gain will be

$$IG(X) = \sum_i U(P(C_i)) - E[\sum_i U(P(C_i|X))]$$

i.e. information gain can be defined as the difference of prior uncertainty and uncertainty after considering feature X.

1.5.2 Distance

Distance [40] measure specifies the discrimination power of feature, i.e. how strongly a feature can discriminate between classes. A feature with higher discrimination power is better than the one with less discrimination power. If Ci and Cj are two classes and X is the feature, then distance measure D(X) will be the difference of P(X|Ci) and P(X|Cj), i.e. difference of probability of 'X' when class is Ci and when class is Cj. X will be preferred over Y if D(X) > D(Y). The larger the difference, the larger means the distance and more preferred the feature is. If P(X|Ci) = P(X|Cj), feature X cannot separate classes Ci and Cj.

1.5.3 Dependency

Instead of the information gain or convergence power, dependency measure determines how strongly two features are connected with each other. In simple words, dependency specifies about how uniquely the values of a feature determine values of other features. In case of supervised learning, it could be dependency of class label 'C' on feature 'X' while in case of unsupervised mode i,t may be dependency of other features on the one under consideration. We may select the features on which other features have high dependency value. If D(X) is dependency of class C on Feature X, then feature X will be preferred over feature Y if D(X) > D(Y).

1.5.4 Consistency

One of the criteria of feature selection may be to select those features that provide the same class structure as provided by entire feature set. The mechanism is called consistency [40], i.e. to select the features with the condition P(C|Subset) = P(C| Entire Feature Set). So we have to select the features that maintain the same consistency as maintained by entire dataset.

1.5.5 Classification Accuracy

Classification accuracy is normally dependent on the classifier used and is suitable for wrapper based approaches. The intention is to select the features that provide the best classification accuracy based on the feedback from the classifier. Although this measure provides quality features (as classification algorithm is also involved), but it has few drawbacks, e.g. How to estimate accuracy and avoid overfitting, noise in the data may lead to wrong accuracy measure. It is computationally expensive to calculate accuracy as classifier takes time to learn from data.

1.6 Feature Generation Schemes

For any feature selection algorithm, the next feature subset generation is the key point, i.e. to select the members of feature subset for next attempt if in the current attempt the selected feature subset does not provide an appropriate solution. For this purpose, we have four basic feature subset generation schemes.

1.6.1 Forward Feature Generation

In forward feature generation scheme, we start with empty feature subset and add features one by one until feature subset meets the required criteria. The maximum size of feature subset selected may be equal to the total number of features in the entire dataset. A generic forward feature generation algorithm might look like the one given in Fig. 1.7.

'X' is the feature set comprising of all the features from dataset. There are total n number of features, so |X| = n, 'Solution' is the final feature subset that will be selected by the algorithm.

In step-a, we have initialized 'Solution' as an empty set, it does not contain any value. Step-c adds the next feature in sequence to the 'Solution'. Step-d evaluates the new feature subset (after adding the latest feature). If the condition is satisfied,

```
Input:
S - data sample with features X,|X| = n
J - evaluation measure
Initialization:
a)  Solution ← {ϕ}
do
b) ∀x ∈ X
                    c) Solution ← Solution ∪ {xᵢ}    i = 1..n
d) until Stop(J, Solution)
Output
d) Return Solution.
```

Fig. 1.7 A sample forward feature generation algorithm

the current feature subset is returned as selected subset otherwise the next feature in the sequence is added to the 'Solution'. The process continues until the stopping criteria are met in which case the 'Solution' is returned.

1.6.2   Backward Feature Generation

Backward feature generation is opposite to forward generation scheme: in forward generation, we add with an empty set and add features one by one. In backward feature generation, on the other hand, we start with full feature set and keep on removing the features one by one, until no further feature can be removed without affecting the specified criteria. a generic backward feature generation algorithm is given in Fig. 1.8.

In backward feature generation algorithm, initially at Step-a, the entire feature set is assigned to 'Solution'. Then we keep on removing features one by one. A new subset is generated after removing a feature. So, $|Solution| \leftarrow |Solution| - 1$. After removing the ith feature, the evaluation condition is checked. A typical condition might be that removing of a feature might not affect the classification accuracy of the remaining features or it might be that 'A feature can be removed if, after

```
Input:
S - data sample with features X,|X| = n
J - evaluation measure
Initialization:
a)  Solution ← X
do
b) ∀x ∈ Solution
                    c) Solution ← Solution − {xᵢ}  i = n..1
d) until Stop(J, Solution)
Output
d) Return Solution.
```

Fig. 1.8 A sample backward feature generation algorithm

removing this feature, consistency of remaining features remain the same as that of entire feature set. Mathematically, if 'R(Solution)' is consistency of a feature subset 'Solution' and Xi is the ith feature in 'Solution' then:

If R(Solution − {Xi}) = R(Entire Feature Set) then

Solution = Solution − {Xi}

Note: We can also use hybrid feature generation approach in which case both forward generation scheme and backward generation scheme are followed concurrently. Two search procedures run in parallel where one search starts with an empty set and other starts with the full set. One adds features one by one while other removes features one by one. The procedure stops when any of the search finds feature subset according to a specified condition.

1.6.3   Random Feature Generation

There is another strategy called random feature generation. In random feature generation apart from all of the above-mentioned strategies, we randomly select features to be part of the solution. Features can be selected or skipped on any specified criteria. A simple procedure may be to select the features on the basis of random number generation mechanism. Following procedure generates a random number between 0 and 1, if number is <0.5, the current feature will be included else it will be excluded.

$$Solution \leftarrow \{\phi\}$$
$$For\, i = 1\, to\, n$$
$$If\, (Rand(0, 1) < 0.5)\, then$$
$$Solution \leftarrow Solution \cup \{X_i\}$$

On the basis of this mechanism, a typical hit-and-trial feature selection algorithm might look like the one given in Fig. 1.9.

Input:
S - data sample with features X,|X| = n
J - evaluation measure

Solution ← {ϕ}
For i=1 to n
If (Rand(0,1) < 0.5) then
Solution ← Solution ∪ {X_i}
Until Stop(J, Solution)
Output
d) Return Solution.

Fig. 1.9 Sample Hit and Trial algorithm based on random feature generation algorithm

Random feature generation approach is followed in Genetic-, Particle Swarm-, Fish- Swarm-like algorithms. However, in such algorithms, some heuristics are also followed to generate the next feature subset along with random generation mechanism. We will discuss on it in upcoming chapters.

1.7   Related Concepts

Now, we will present some important preliminary concepts regarding features selection.

1.7.1   Search Organization

For dataset d comprising of features X, a search algorithm needs to explore the feature space. The most intuitive method is to search the entire dataset and find a candidate subset [6]. Unfortunately, exhaustive search is not possible for datasets beyond smaller size and leads to expensive computational space. Hence exhaustive search is only suitable for datasets with smaller size.

An alternate way is to use random search [40], where a random set of features is selected and evaluated against their capability of being the required feature subset. The process continues until we get a potential candidate solution is found or a predefined time period has elapsed. Third and more commonly used methods are to imply heuristic-based search [40] for feature subset selection. In these methods, a heuristic function is used to guide the search. The search is directed to achieve the maximum value of function. The process continues until we get a solution with the required value of the heuristic function.

1.7.2   Generation of a Feature Selection Algorithm

Based on the above discussion, it is clear that feature selection algorithm should have following three ingredients: Feature Subset Generation component to generate next subset, evaluation measure to evaluate the quality of current subset and search organization to let the algorithm proceed on its way to perform feature selection.

Figure 1.10 is a modified form of characteristics space of feature selection algorithm taken from [37].

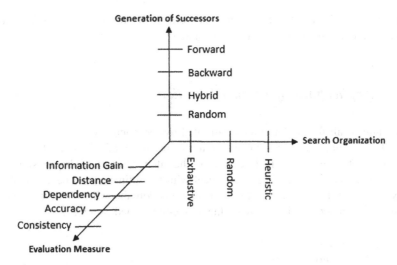

Fig. 1.10 Characteristics space of feature selection algorithm

1.7.3   Feature Relevance

Relevancy is defined in terms of the evaluation measure under consideration. A feature will be irrelevant if removing this feature does not affect the evaluation measure. For example, if D(X) is the dependency of class label on attribute set 'X' then attribute $x_i \in X$ is irrelevant if $D(X - x_i) = D(X)$, i.e. removing the feature x_i does not affect the dependency of class label on the remaining feature set. Similarly, it can be defined for other evaluation measures. Kohavi et al. [41] define the degree of relevance by stating three categories, strongly relevant, weekly relevant and irrelevant. Strongly relevant features are those which cannot be removed without distorting the underlying evaluation measure. Irrelevant features are those which can be safely be removed without distorting the evaluation measure, and those that fall in between (e.g. features having dependency less than strongly relevant features and greater than zero, dependency of irrelevant features is considered to be zero) will be weekly relevant.

1.7.4   Feature Redundancy

Redundant features are those which do not add any information to feature subset. Just like irrelevant features, removal of which does not destroy the evaluation measure, redundant features are the ones adding which do not improve evaluation measure. If D(X) is the dependency of feature subset 'X' then the feature Xj will be redundant if $D(X \cup \{x_j\}) = D(X)$. In other words if two features X and Y have sthe ame dependency, then one of them is redundant.

1.7.5   Applications of Feature Selection

Today almost all fields of life are confronted by the dilemma of high dimensionality. This is not the only problem, but the presence of noisy, irrelevant and redundant data makes it difficult to perform the intended analysis. In this situation, feature selection is an effective tool to deal with high dimensional data to make it ready for further tasks. We will discuss here a few domains where feature selection has played a major role in removing the above-mentioned problems.

1.7.5.1   Text Mining

With the passage of time and emergence of latest technologies day by day, we are overburdened with text, e.g. emails, blogs, books. This requires proper processing of documents, e.g. to build vocabularies, to find a conceptual depth of document regarding a particular concept (e.g. if a word 'Cricket' is repeated 15 times than any other word, then most probably the document explains something about 'Cricket'), to group similar books. Feature selection methods have been successfully applied both for text categorization and clustering. The authors in [44] provide five empirical evidence that feature selection method can improve the efficiency and performance of text clustering algorithm. A number of algorithms have been proposed in literature to effectively use feature selection in text mining, [45–48] are few to name.

1.7.5.2   Image Processing

Representing images is not a straightforward task, as the number of possible image features is practically unlimited [49]. Be it simple surface image or domain of medical imaging, feature selection has played a major role in dimensionality reduction. It can be used for processing/preprocessing of both images and video streams where it can help isolate a portion (features) of image for particular analysis purpose. Few applications of feature selection in medical image processing include image recognition, i.e. identification of important parts from the images in presence of noise, images classification for retrieving images from large repositories on the basis of contents of images, for training purpose, e.g. to automatically distinguish between images of healthy and diseased optic nerves, image cleansing to remove noise for further diagnostics.

1.7.5.3   Bioinformatics

The domain of bioinformatics is normally characterized by large number of input features, e.g. there are millions of single-nucleotide polymorphisms (SNPs), thousands of genes in microarrays. So, an important task is to find out useful features (e.g. SNPs, genes) in various situations, e.g. to classify a certain community from other or to diagnose a disease. Feature selection is an important tool that has successfully been deployed for this purpose. It helps to focus on important or useful features only while ignoring the rest. It can be applied on microarray analysis, genomic analysis, mass spectra analysis, etc.

1.7.5.4   Intrusion Detection

With the passage of time, Internet has become a must-to-have utility for every person. Although getting benefits from it, one is always at risk of any type of intrusion while using Internet. On daily basis, intruders are coming with new ways to compromise your security. The static measures such as firewalls, antivirus applications are not necessary, we need to have more dynamic intrusion detection systems (IDS), IDSs can be both host based that monitor the local system resources, e.g. files, logs, disks, whereas network-based systems monitor network traffic. However, the problem with these systems is the huge amount of data passed through the network. Here, we can use feature selection process to detect important features of an event and then train the system on these features to automatically detect any intrusion in future.

Apart from above-mentioned domains, there are various other domains where feature selection has been successfully used, e.g. business and finance, industries, weather forecasting, remote sensing, network communication. Covering all of these domains is out of scope of the book, so only a few domains and use of feature selection in these domains have been discussed.

1.7.6   Feature Selection: Issues

Since its commencement, lot of improvement has been made in feature selection process, however, there are various issues that are still challenging. We will discuss a few issues here.

1.7.6.1   Scalability

The size of data is growing day by day, the increase is found both in number of instances and number of features. It's common to have real-world applications with thousands of features and instances. The size has grown to the extent that it is difficult to fit the datasets in memory thus has posed a great challenge for feature selection algorithms to cater scalability of datasets. Majority of feature selection algorithms need the entire dataset to keep in memory for exact calculation, which may demand more number of resources because it is hard for the algorithms to calculate ranks/relevance on the basis of less number of samples or alternatively calculate using different passes, which may affect the performance of these algorithms. So, the conclusion is that with rapidly growing size of datasets, it has become critical to address the scalability of feature selection algorithms as well.

1.7.6.2 Stability

Stability of feature selection algorithms is another challenge the community is focused on, feature selection algorithms may produce different features in case of small amount perturbation. Whereas they are required to produce the same results, i.e. they should be stable. There are various factors that can affect the output of a specific feature selection algorithm and thus affect its sensitivity, e.g. dimensionality m, sample size n and different data distribution across different folds [39].

1.7.6.3 Linked Data

Almost all of the feature selection algorithms, assume that data is independent and identically distributed, however, the important factor that is ignored is that data may be linked as well. Especially with the emergence of social media like Facebook, Twitter, where instances are linked with each other (users linked with posts, posts liked by other users, users make tweets, tweets are retweeted by other users, etc.). Unfortunately, the task of feature selection for linked data is rarely touched. The major challenges faced by feature selection algorithms in case of linked data are [39] the relations between the linked data and how to use these relations for feature selection. Although work has already been started in feature selection for linked data, e.g. [42, 43], however, it is still a challenge that feature selection community needs to consider on a priority basis.

1.8 Summary

In this chapter, we have discussed the required preliminary concepts of feature selection. We started from the very basic concepts of features to different feature selection methods to underlying techniques to its applications and challenges. Normally, the process of feature selection as explained in other books and tutorials comes with heavy mathematical equations, algebraic and statistical concepts which itself is a bigger challenge for a researcher especially the newcomer in this domain to understand. Efforts are made to explain the concepts in a very simple way and with examples where possible. We also provided the generic pseudocode of algorithms to give the idea of how a particular type of algorithm might work. This will be a special help for the readers to understand more advanced algorithm and develop their own.

References

1. Bishop CM (2006) Pattern recognition and machine learning, vol 128, pp 1–58
2. Domingos P (2012) A few useful things to know about machine learning. Commun ACM 55 (10):78–87
3. Villars RL, Olofson CW (2011) Big data: what it is and why you should care. White Paper, IDC 14
4. Jothi N, Husain W (2015) Data mining in healthcare—a review. Proc Comput Sci 72:306–313
5. Kaisler S et al (2013) Big data: issues and challenges moving forward. In 2013 46th Hawaii international conference on system sciences (HICSS). IEEE
6. Jensen R, Shen Q (2008) Computational intelligence and feature selection: rough and fuzzy approaches, vol 8. Wiley
7. Bellman R (1956) Dynamic programming and lagrange multipliers. Proc Natl Acad Sci 42 (10):767–769
8. Neeman S (2008) Introduction to wavelets and principal components analysis. VDM Verlag Dr. Muller Aktiengesellschaft & Co, KG
9. Engelen S, Hubert M, Branden KV (2016) A comparison of three procedures for robust PCA in high dimensions. Austrian J Stat 34(2):117–126
10. Cunningham P (2008) Dimension reduction. In: Machine learning techniques for multimedia. Cognitive Technologies. Springer, Berlin
11. Van Der Maaten L, Postma E, Van den Herik J (2009) Dimensionality reduction: a comparative. J Mach Learn Res 10:66–71
12. Friedman JH, Stuetzle W (1981) Projection pursuit regression. J Am Stat Assoc 76(376):817–823
13. Borg I, Groenen P (2005) Modern multidimensional scaling: theory and applications. Springer Science & Business Media
14. Dalgaard P (2008) Introductory statistics with R. Springer Science & Business Media
15. Zeng X, Luo S (2008) Generalized locally linear embedding based on local reconstruction similarity. In: Fifth international conference on fuzzy systems and knowledge discovery, FSKD'08, vol. 5. IEEE
16. Saul LK et al (2006) Spectral methods for dimensionality reduction. Semisupervised Learn 293–308
17. Liu R et al (2008) Semi-supervised learning by locally linear embedding in kernel space. In: 19th international conference on pattern recognition, ICPR 2008. IEEE
18. Gerber S, Tasdizen T, Whitaker R (2007) Robust non-linear dimensionality reduction using successive 1-dimensional Laplacian eigenmaps. In: Proceedings of the 24th international conference on machine learning. ACM
19. Donoho DL, Grimes C (2003) Hessian eigenmaps: locally linear embedding techniques for high-dimensional data. Proc Natl Acad Sci 100(10):5591–5596
20. Teng L et al (2005) Dimension reduction of microarray data based on local tangent space alignment. In: Fourth IEEE conference on cognitive informatics, (ICCI 2005). IEEE
21. Raman B, Ioerger TR (2002) Instance-based filter for feature selection. J Mach Learn Res 1 (3):1–23

22. Yan G et al (2008) Unsupervised sequential forward dimensionality reduction based on fractal. In: Fifth international conference on fuzzy systems and knowledge discovery, FSKD'08, vol 2. IEEE
23. Tan F et al (2008) A genetic algorithm-based method for feature subset selection. Soft Comput 12(2):111–120
24. Loughrey J, Cunningham P (2005) Using early-stopping to avoid overfitting in wrapper-based feature selection employing stochastic search. In: Proceedings of the twenty-fifth SGAI international conference on innovative techniques and applications of artificial intelligence
25. Valko M, Marques NC, Castellani M (2005) Evolutionary feature selection for spiking neural network pattern classifiers. In: 2005 Portuguese conference on artificial intelligence. IEEE
26. Huang J, Lv N, Li W (2006) A novel feature selection approach by hybrid genetic algorithm. In: Trends in artificial intelligence, PRICAI 2006, pp 721–729
27. Khushaba RN, Al-Ani A, Al-Jumaily A (2008) Differential evolution based feature subset selection. In: 2008 19th international conference on pattern recognition, ICPR 2008. IEEE
28. Roy K, Bhattacharya P (2008) Improving features subset selection using genetic algorithms for iris recognition. In: IAPR workshop on artificial neural networks in pattern recognition. Springer, Berlin
29. Dy, Jennifer G., and Carla E (2004) Brodley. Feature selection for unsupervised learning. J Mach Learn Res 5:845–889
30. He X, Cai D, Niyogi P (2005) Laplacian score for feature selection. In: NIPS, vol 186
31. Wolf L, Shashua A (2005) Feature selection for unsupervised and supervised inference: the emergence of sparsity in a weight-based approach. J Mach Learn Res 6:1855–1887
32. Bryan K, Cunningham P, Bolshakova N (2005) Biclustering of expression data using simulated annealing. In: 2005 Proceedings 18th IEEE symposium on computer-based medical systems. IEEE
33. Handl J, Knowles J, Kell DB (2005) Computational cluster validation in post-genomic data analysis. Bioinformatics 21(15):3201–3212
34. Gluck M (1985) Information, uncertainty and the utility of categories. In: Proceedings of the seventh annual conference on cognitive science society. Lawrence, Erlbaum
35. Dash M, Liu H (1997) Feature selection for classification. Intell Data Anal 1(1-4):131–156
36. Vanaja S, Ramesh Kumar K (2014) Analysis of feature selection algorithms on classification: a survey. Int J Comput Appl 96(17)
37. Ladha L, Deepa T (2011) Feature selection methods and algorithms. Int J Comput Sci Eng 3(5):1787–1797
38. John GH, Kohavi R, Pfleger K (1994) Irrelevant features and the subset selection problem. In: Proceedings of the eleventh international conference on machine learning
39. Tang J, Alelyani S, Liu H (2014) Feature selection for classification: a review. Data Classif Algorithms Appl 37
40. Hua J, Tembe WD, Dougherty ER (2009) Performance of feature-selection methods in the classification of high-dimension data. Pattern Recogn 42(3):409–424
41. Kohavi R, John GH (1997) Wrappers for feature subset selection. Artif Intell 97(1–2):273–324
42. Tang J, Liu H (2012) Feature selection with linked data in social media. In: Proceedings of the 2012 SIAM international conference on data mining. Society for industrial and applied mathematics
43. Gu Q, Han J (2011) Towards feature selection in network. In: Proceedings of the 20th ACM international conference on information and knowledge management. ACM
44. Liu T et al (2003) An evaluation on feature selection for text clustering. In: ICML, vol 3
45. Tutkan M, Ganiz MC, Akyokuş S (2016) Helmholtz principle based supervised and unsupervised feature selection methods for text mining. Inf Process Manag 52(5):885–910
46. Özgür L, Güngör T (2016) Two-stage feature selection for text classification. In: Information sciences and systems, pp 329–337. Springer International Publishing

47. Liu M, Xiaoling L, Song J (2016) A new feature selection method for text categorization of customer reviews. Commun Stat-Simul Comput 45(4):1397–1409
48. Kumar V, Sonajharia M (2014) Multi-view ensemble learning for poem data classification using SentiWordNet. In: Advanced computing, networking and informatics-volume 1, pp 57–66. Springer International Publishing
49. Bins J, Draper BA (2001) Feature selection from huge feature sets. In: 2001 Proceedings of eighth IEEE international conference on computer vision, ICCV 2001, vol 2. IEEE

Chapter 2
Background

To overcome the phenomenon of curse of dimensionality, one of the methods is to reduce dimensions without effecting relevant information present in entire dataset. There are various techniques proposed in the literature to reduce dimensions. In this chapter, we have presented an overview of these techniques.

2.1 Curse of Dimensionality

Data creation is occurring at a record rate [1]. This increase is seen in all areas of human activity right from the data generated on daily basis such as telephone calls, bank transactions, business tractions to more technical and complex data including astronomical data, genome data, molecular datasets, medical records, etc. These datasets may contain lot of information useful but still undiscovered. This increase in data is twofold, i.e. both in number of samples/instances and number of features that are recorded and calculated. As a result, many real-world applications need to process datasets with hundreds and thousands of attributes. Few of such datasets are also publically available at [2].

The significant increase in the number of dimensions in datasets leads to phenomenon called *curse of dimensionality*. The curse of dimensionality is the problem caused by the exponential increase in volume associated with adding extra dimensions to a (mathematical) space [4]. Dimension reduction (DR) is used as preprocessing [13]. The original feature space is mapped onto a new, reduced dimensionality space and the samples are represented in that new space [5]. Normally datasets contain number of misleading and redundant information which needs to be removed before any further tasks can be performed on these datasets. For example, in case of deriving complex classification rules, it is more effective to first perform dimension reduction. This step not only enhances the performance but also increases the resulting classification accuracy and making the rules more comprehensible.

© Springer Nature Singapore Pte Ltd. 2019   27
M. S. Raza and U. Qamar, *Understanding and Using Rough Set
Based Feature Selection: Concepts, Techniques and Applications*,
https://doi.org/10.1007/978-981-32-9166-9_2

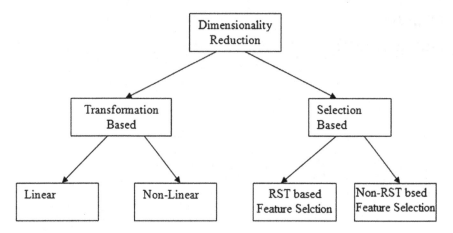

Fig. 2.1 Taxonomy of dimensionality reduction

There are various techniques to perform DR, e.g. [6–10], but many of such techniques destroy the underlying semantics of data which makes them undesirable to many real-world applications.

So, primarily, this book will focus on DR techniques that preserve original data semantics. In particular, we will focus on those techniques based on Rough Set Theory [11]. Taxonomy of DR techniques is presented in Fig. 2.1. The presented techniques are classified into two categories: those that change the underlying semantics of data during DR process and those that preserve data semantics. The choice of the method is dependent upon the underlying application, e.g. if an application needs to preserve original data semantics than the DR technique to be chosen ensure that it is preserved. However, if an application requires to discuss the relationships between attributes then the techniques that transforms the data into two or three dimensions while emphasizing these relationships may be selected.

Selection-based techniques can be both RST-based or Non-RST-based. Besides these, there are other techniques which perform semantics-preserving dimensionality in sideline, e.g. machine learning algorithm C4.5 [12]. In this chapter, we will discuss sample techniques from each of the above categories. Such techniques are out of scope of this book.

2.2 Transformation-Based Reduction

These are one of the common techniques in DR literature. They perform DR process but as sideline transform the descriptive dataset features. These techniques are useful where the semantics of original features are not needed by any future process. This section discusses some of such techniques, these are classified into two categories: linear and non-linear.

2.2.1   Linear Methods

Various Linear methods for DR are proposed in literature with passage of time and include techniques like Principal Component Analysis [13–17] and Multidimensional Scaling [18].

2.2.1.1   Principal Component Analysis (PCA)

Principal component analysis (PCA) [13, 14] is a well-known tool for data analysis and transformation, and is considered the canonical means of DR. PCA is a mathematical tool that converts large number of correlated variables to smaller number of uncorrelated variables called components. The intention is to reduce the dimensions in dataset but still preserving original variability in data. The first principal component accounts for maximum of variability possible and each of the succeeding component accounts for maximum of remaining variability.

PCA represents variance–covariance structure of high dimensional vector with few linear combinations of the original component variables. For example, for a p-dimensional random vector $\underline{X} = (X_1, X_2, ..., X_p)$, PCA will find k (univariate) random variables $Y_1, Y_2, ..., Y_k$ called K principal components and can be defined by the following formula:

$$Y_1 = l'_1\underline{X} = l_{11}X_1 + l_{12}X_2 + \ldots + l_{1p}X_P$$
$$Y_2 = l'_2\underline{X} = l_{21}X_1 + l_{22}X_2 + \ldots + l_{2p}X_P$$
$$\vdots$$
$$Y_k = l'_k\underline{X} = l_{k1}X_1 + l_{k1}X_2 + \ldots + l_{pk}X_P$$

Here, $l_1, l_2,...$ etc. coefficient vectors which are chosen on the basis of the following conditions:

- First Principal Component = Linear combination $l_1\,'\underline{X}$ that maximizes Var $(l_1\,'\underline{X})$ and $\|l_1\| = 1$.
- Second Principal Component = Linear combination $l_2\,'\underline{X}$ that maximizes Var $(l_2\,'\underline{X})$ and $\|l_2\| = 1$ and $\text{Cov}(l_1\,'\underline{X}, l_2\,'\underline{X}) = 0$.
- j th Principal Component = Linear combination $l_j\,'\underline{X}$ that maximizes Var $(l_j\,'\underline{X})$ and $\|l_j\| = 1$ and $\text{Cov}(l_k\,'\underline{X}, l_j\,'\underline{X}) = 0$ for all k < j.

It means that each principal component is a linear combination that maximizes the variance of linear combination and has zero covariance with the previous component. Figure 2.2 shows the two-dimensional normal point cloud with corresponding principal components.

Thus, PCA maximizes the variance of datasets sample vectors along their axes by locating a new coordinate system and suitably transforms the samples. The new axes are constructed in decreasing order of variance such that the first variable in new axes has maximum variance and so on. Correlation in new sample space is

Fig. 2.2 Two-dimensional
normal point cloud with
corresponding principal
components [19]

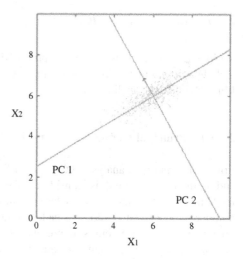

reduced or totally removed consequently resulting in reduced redundancies. Thus, DR can be performed on a dataset using PCA and then selecting appropriate number of first k principal components as per requirement and discarding the rest.

PCA, however, suffers from the following shortcomings:

- It destroys the underlying semantics of data.
- It can be used only for numeric datasets.
- It can only deal with linear projects and thus ignores any non-linear structure in the data.
- Finally, human input is also required to decide how many of the first principal components will be kept. Thus, the operator's task is to balance information loss against DR to suit the task at hand.

2.2.1.2   Projection Pursuit

Projection pursuit (PP) [20, 21] uses a quality matrix in order to project the data to lower dimensions it opts to select interesting projects by using local optimization over projection directions of a certain 'interestingness' index.

It finds the projections from high-to-low-level projection that reveal maximum of the information about the structure of the original dataset. After finding the required projections, the existing structures (clusters, surfaces, etc.) can be extracted and analysed separately.

The most basic form of PP is the scatterplot [21]. Scatterplot in its simplest form uses two dimensions to display data characteristics at a time. It is very easy to produce all two-dimensional scatterplots out of n dimensions to perform analysis. Totally, there can be (2^n) pair-wise scatterplots. However, this only allows the tasks to be performed only using two-dimensional scatterplots.

Hence, the projections that give single-dimensional projects distribution are considered to be interesting far from normal distributions. We can use different projection indices arising out of different forms of normality deviations. Friedman and Tukey in [21] opted to automate the task of projection pursuit. Numerical indices were used to describe the amount of structure presented in a projection. The heuristic search can then be applied to find the 'interesting' projections using these indices. After finding a structure, it is removed from the data and data is examined for further structures and process continues until there is no further structure to discover in the data.

Different projections may be found, each highlighting a different aspect of the structure of high dimensional data. Typically, linear projections are used due to their simplicity and interpretability.

PP has some disadvantages as well as that of PCA. PP is used for linear data so not appropriate for non-linear one.

2.2.1.3 Multidimensional Scaling

Multidimensional scaling (MDS) [18, 22] converts high-dimensional data to low dimensions using the distance between data points. So, these methods are also called distance methods, the distance, however, can be measured in terms of similarity or dissimilarity measure between data points. MDS, thus converts high-dimensional data to low-dimensional data (normally, these dimensions can be two to three) and tries to preserve interpoints distances up to maximum [23].

Visually, MDS maps the high dimensional data to low-dimensional data such that the patterns in the original dataset are very much present in the converted lower dimensional space. Visually, if we draw the original space and the converted space, the points closer to each other in original space, are also close to each other in converted space, similarly, the points away from each other will remain away in converted space. For example, for a given set of different brands of air fresheners, if two, brands are close to each other, then MDS maps them such that they remain close to each other on converted maps.

From a slightly more technical point of view, what MDS does is find a set of vectors in p-dimensional space such that the matrix of Euclidean distances among them corresponds as closely as possible to some function of the input matrix according to a criterion function called *stress* [22].

A simplified view of the algorithm is as follows:

1. Assign points to arbitrary coordinates in p-dimensional space.
2. Compute Euclidean distances among all pairs of points, to form the matrix.
3. Compare the matrix with the input D matrix by evaluating the stress function The smaller the value, the greater the correspondence between the two.
4. Adjust coordinates of each point in the direction that best maximally stress.
5. Repeat Steps 2 through 4 until stress won't get any lower.

As well as Euclidean distance is concerned, both PCA and MDS are equivalent. However, with MDS, we can use various other methods with different types of metrics and calculations. Irrespective of which method is used, basic principal remains the same. The starting point for MDS is the determination of the 'spatial distance model' [22]. Following are some notations used to determine the proximities between data points.

Let Δ and D are two matrices of $N \times N$ dimension representing different collection of objects. The objects in the matrices are indexed by i and j, where δij is the proximity or data value of object i with object j [22], and dij represents the distance between pairs of points xi and xj as shown in the below equations:

$$D = \begin{bmatrix} \delta_{11} & \delta_{12} & \cdots & \delta_{1N} \\ \delta_{21} & \delta_{22} & \cdots & \delta_{2N} \\ \vdots & \ddots & \vdots & \vdots \\ \delta_{N1} & \delta_{N12} & \cdots & \delta_{NN} \end{bmatrix}$$

$$D = \begin{bmatrix} d_{11} & d_{12} & \cdots & d_{1N} \\ d_{21} & d_{22} & \cdots & d_{2N} \\ \vdots & \ddots & \vdots & \vdots \\ d_{N1} & d_{N12} & \cdots & d_{NN} \end{bmatrix}$$

The aim of MDS is to find a configuration such that the distances dij match, as closely as possible, the dissimilarities δij [22].

We can use different variations of MDS based on functions that transform the proximities. Classical Metric Multidimensional Scaling is a basic form of MDS where dij are as close as possible to δi using Euclidean distance. This is also sometimes referred to as *principal coordinate analysis*, which is also equivalent to PCA [23]. In Classical MDS methods, the relationships between δij and dij depends upon the metric properties of dissimilarities. Nonmetric MDS, on the other hand, refer to those methods where the relationship between δij and dij depend on the rank ordering of the dissimilarities [22].

In its original form, MDS [22] used the distance between datapoints and sought for the configuration that would give the similar approximation. Often, a linear projection onto a subspace obtained with PCA is used. The basic idea of MDS (i.e. to approximate the distance between points into new low-dimensional subspace), however, can also be used for constructing a non-linear projection method.

2.2.2 Non-linear Methods

Linear DR methods are no doubt useful but their utility fails in case of non-linear data. This motivated the development of non-linear DR methods, e.g. [24–26]. An example of non-linear method is locally linear embedding (LLE) [27, 28].

2.2.2.1   Locally Linear Embedding

LLE calculates reconstructions (embedding) which are low dimensional and neighbourhood preserving by using local symmetries of linear reconstructions (from high dimensional data). This can be explained better by considering the following informal analogy [28]. Initial data is three dimensional, however, taking shape of rectangular manifold (two dimensional) that has been moulded to a three dimensional S-shaped curve. Now, Scissors cut this manifold into small squares. Each square represents a locally linear patch of the non-linear surface. These squares are then arranged on flat surface, however, by preserving angular relationships between neighbouring squares. As all transformations comprise of translation, scaling or rotation only so this is a linear mapping. Through this process algorithm uses series of linear steps to find non-linear structure.

In first step, it selects neighbours in data points. This selection can be achieved using Euclidean distance for k nearest neighbours [29]. In the second step, LLE computes the weights that linearly reconstruct data points using least square problem. The following cost function is used:

$$E_1(W) = \sum_{i=1}^{N} \left| (X_i - \sum_{j=1}^{k} W_{ij} X_{Nj}) \right|^2 \tag{2.1}$$

Finally, we compute low-dimensional embedded vectors by minimizing the embedded cost function:

$$E_2(Y) = \sum_{i=1}^{N} \left| (Y_i - \sum_{j=1}^{k} W_{ij} Y_{Nj}) \right|^2 \tag{2.2}$$

Figure 2.3 summarizes LLE algorithm.
Figure 2.3 gives the overview of these three steps.
To save the time and space, LLE also tends to accumulate very sparse matrices. It avoids dynamic programming problems as well. LLE, however, does not provide any indication about how to map a test data point from input space to manifold space or how to reconstruct a data point from its low-dimensional representation.

---

1. Compute the neighbours of each data point, $\vec{X}_i$.
2. Compute the weights W_{ij} that best reconstruct each data point $\vec{X}_i$ from its neighbours, minimizing the cost in eq. (1) by constrained linear fits.
3. Compute the vectors $\vec{Y}_i$ best reconstructed by the weights W_{ij}, minimizing the quadratic form in eq. (2) by its bottom nonzero eigenvectors.

Fig. 2.3   Summary of the LLE algorithm

Similar to LLE, Laplacian Eigenmaps attempts to find low-dimensional data representation while preserving local properties of the manifold [30]. In Laplacian Eigenmaps, the local properties are based on neighbours. Laplacian Eigenmaps minimizes the distance between a data point and its k nearest neighbours in an attempt to construct low-dimensional representation of the data. Weights are used for this purpose, i.e. the distance between a datapoint and its first nearest neighbour contributes more to the cost function as compared to the distance between the datapoint and its second nearest neighbour, which costs more as compared to distance between datapoint and its third neighbour and so on. Using spectral graph theory, the minimization of the cost function is defined as an Eigen problem.

2.2.2.2 Isomap

Isomap [31] uses geodesic interpoint distances in contrast to Euclidean distances. The geodesic distance between the two red points is the length of the geodesic path, which is the shortest path between the points that lies on the surface [32]. Figure 2.4 shows the geodesic distance between two red points.

Isomap deals with finite datasets of points in R^n which are assumed to lie on a smooth submanifold M^d of low dimension $d < n$. From the given datapoints, algorithm tries to recover M. For a given graph G constructed out of the datapoints, Isomap attempts to find geodesic distance between data points in M. The Isomap algorithm consists of the following three basic steps:

(1) Determine which points are neighbours on the manifold M, based on the distances between pairs of points in the input space.
(2) Estimate the geodesic distances between all pairs of points on the manifold M by computing their shortest path distances in the graph G.
(3) Apply MDS to matrix of graph distances, constructing an embedding of the data in a d-dimensional Euclidean space Y that best preserves the manifold's estimated geometry.

Fig. 2.4 Geodesic distance between two red points

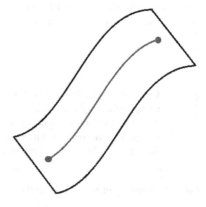

Selection of 'Neighbourhood' size in Isomap is critical because a large size may introduce 'short-circuit' into neighbourhood graph whereas small size may make graph too sparse to approximate geodesic distance.

The success of Isomap depends on being able to choose a neighbourhood size (either ε or K) that is neither so large that it introduces 'short-circuit' edges into the neighbourhood graph, nor so small that the graph becomes too sparse to approximate geodesic paths accurately. Short-circuit edges occur where there are links between data points that are not near each other geodesically and can lead to low-dimensional embeddings that do not preserve a manifold's true topology. The choice of neighbourhood size is an obvious limitation [31].

2.2.2.3 Multivariate Adaptive Regression Splines (MARS)

MARS [34] intends for solving regression type problems for predicting the value of decision class from a set of conditional features. Without making any assumption about underlying functional relationships it constructs this relation using basis functions:

The basis function may be of the form

$$(x - t)_+ = \begin{cases} x - t & x > t \\ 0 & otherwise \end{cases}$$

t is the control point of a basis function and is determined from data.
The general MARS algorithm is as follows:

1. Begin with the simplest model involving only the constant basis function.
2. For each variable and possible control points, search basis functions space and add them according to certain criteria (e.g. minimisation of the prediction error).
3. Repeat Step-2 until a model of predetermined maximum complexity is obtained.
4. Apply the pruning procedure by removing the basis functions that contribute least to the overall (least squares) goodness of fit.

Brute force approach is applied to find variables, interactions and control points, whereas least square procedure is used to determine regression coefficients. However, MARS may also generate complex mode due to noise in the data.

2.3 Selection-Based Reduction

In contrast to transformation-based techniques, which destroy the underlying semantics of data, semantics-preserving DR techniques (called feature selection) preserve original data semantics. The main intention of feature selection is to find minimal feature subset from entire dataset while maintaining a high accuracy in representing original features. Data in real work may be noisy, containing irrelevant and misleading features so, many real-world applications require FS to fix the problem. For instance, by removing these factors, learning from data techniques can benefit greatly.

2.3.1 Feature Selection in Supervised Learning

In supervised learning, feature subset selection explores feature space, generates candidate subsets and evaluates/rates them on the basis of criterion, which serves as a guide to search process. The usefulness of a feature or feature subset is determined by both its *relevancy* and *redundancy* [2]. A feature is relevant if it determines the value of decision feature(s), otherwise, it will be irrelevant. A redundant feature is the one highly correlated with other features. Thus, a good feature subset is the one highly correlated with decision feature(s) but uncorrelated with each other.

The evaluation schemes used in both supervised and unsupervised feature selection techniques can generally be divided into three broad categories [35, 36]:

1. Filter approaches.
2. Wrapper methods.
3. Embedded approaches.

2.3.2 Filter Techniques

Filter techniques perform feature selection independent of learning algorithms. Features are selected on the basis of some rank or score. A score indicating the 'importance' of the term is assigned to each individual feature based on an independent evaluation criterion, such as distance measure, entropy measure, dependency measure and consistency measure [37]. Various feature filter-based feature selection techniques have been proposed in literature, e.g. [38–40]. In this section, we will discuss some representative filter techniques along with advantages and disadvantages of each.

2.3.2.1 Focus

FOCUS [41] uses breadth-first search to find feature subsets that give consistent labelling of training data. It evaluates all the subsets of current size (initially one) and removes ones with least one inconsistency. The process continues until it finds a consistent subset or has evaluated all the possible subsets. Algorithm, however,

Fig. 2.5 Pseudocode of
FOCUS algorithm

FOCUS(O, c).
O, the set of all objects;
c, the number of conditional features;

(1) $R \leftarrow \{\}$
(2) **for** num = 1...c
(3) **for** each subset L of size num
(4) cons = determineConsistency(L, O)
(5) **if** cons == true
(6) $R \leftarrow L$
(7) **return** R
(8) **else** continue

suffers from two major drawbacks: it is very sensitive to noise or inconsistencies in training datasets and algorithm furthermore, due to exponential growth of the features power set size, algorithm is not suitable for application in domains having large number of dimensions. Figure 2.5 shows the pseudocode of FOCUS algorithm

2.3.2.2 Relief

RELIEF [42] works by assigning the relevance weights to each attribute based on its ability to discern objects between different decision classes. It randomly selects an objects, finds its nearHit (objects with same class labels) and nearMiss (i.e. the objects with different class labels). The distance between two objects is the sum of the number of features that differ in value between them:

$$dist(o, x) = \sum_{i=1}^{|C|} diff(o_i, x_i) \tag{2.3}$$

where

$$diff(o_i, x_i) = \begin{cases} 1, & o_i \neq x_i \\ 0, & o_i = x_i \end{cases} \tag{2.4}$$

i.e. the distance between objects is '1' if value of an attribute differ between them and '0' if value of an attribute is the same for both objects. Algorithm requires manual threshold value which should specify that which attribute(s) will finally be selected. The algorithm, however, fails to remove redundant features as two predictive but highly correlated features are both likely to be given high relevance weightings. Figure 2.6 shows the pseudocode of RELIEF algorithm.

```
RELIEF(O, c, S, ε).
O, the set of all objects;
c, the number of conditional features;
S, the number of iterations;
ε, weight threshold value.
    (1) R ← { }
    (2) ∀ Wₐ, Wₐ ← 0
    (3) for i = 1 … S
    (4) choose an object x in O randomly
    (5) calculate x's nearHit and nearMiss
    (6) for j = 1...c
    (7) Wⱼ ← Wⱼ − diff(Xⱼ, nearHitⱼ)/itS + diff(Xⱼ, nearMissⱼ)/itS
    (8) for j = 1...c
    (9) if Wⱼ ≥ ε; R ← R ∪ {j}
    (10)        return R
```

Fig. 2.6 Pseudocode of RELIEF algorithm

2.3.2.3 Selection Construction Ranking Using Attribute Pattern (SCRAP)

Figure 2.7 shows the pseudocode of SCRAP [43] algorithm.

It (SCRAP [43]) performs sequential search to determine feature relevance in instance space. It attempts to identify those features that change decision boundaries in dataset by considering one object (instance) at a time, these features are considered to be most informative. Algorithm starts by selecting a random object, which is considered as the first point of class change (PoC). It then selects next PoC, which usually is the nearest object having different class labels. After this, the

```
SCRAP(O).
O, the set of all objects;

    (1) A ← { }; ∀ Wᵢ, Wᵢ = 0;
    (2) T ← randomObject( ); PoC ← T
    (3) while O ≠ { }
    (4) O ← O − PoC; PoC_new ← NewPoC(PoC)
    (5) n = dist(PoC, PoC_new)
    (6) if n == 1
    (7) i = diffFeature(PoC, X); A ← A ∪ {i}
    (8) N ← getClosestNeighbours(PoC, n)
    (9) ∀ X ε N
    (10)        if classLabel(X) == classLabel(N)
    (11)        O ← O − X
    (12)        if dist(PoC, X) == 1
    (13)        i = diffFeature(PoC, X); Wᵢ = Wᵢ − 1
    (14)        else if dist(PoC, X) > 1
    (15)        incrementDifferingFeatures(X, W)
    (16)        R ← A
    (17)        ∀ Wᵢ, if Wᵢ > 0 then R ← R ∪ {i}
```

Fig. 2.7 Pseudocode of SCRAP algorithm

nearest object to this having a different class label which becomes the next PoC. These two PoCs define a neighbourhood and dimensionality of decision boundary between the two classes are defined by the features that change between them. If only one feature changes between them, then it is considered to be absolutely relevant and is included in feature subset otherwise their associated relevance weights (which initially are zero), are incremented. However, if objects in the same class are closer than this new PoC and differ only by one feature then relevance weight is decremented. Objects belonging to neighbourhood are then removed and this process continues until there is no unassigned object to any neighbourhood. Final feature subset is then selected comprising of features with positive relevance weight and those that are absolutely relevant.

Major deficiency of the approach is that it regularly chooses a large number of features. This normally happens in case when weights are decremented. Feature weights remain unaffected if more than one features change between a PoC and an object belonging to the same class.

2.3.3   Wrapper Techniques

One of the criticisms suffered by filter approaches is that the filter to select attributes is independent of the learning algorithm. To overcome this issue, wrapper approaches use classifier performance to guide the search, i.e. the classifier is wrapped in the feature selection process [47]. So, in these approaches, feature subset is selected on the basis of generalization accuracy it offers as estimated using cross-validation on the training data.

Four popular strategies are [35, 44] as follows:

1. Forward selection (FS): Starting with an empty feature subset, it evaluates all features one by one, selects the best feature and combines this feature with others one by one.
2. Backward elimination (BE): Initially it selects all features, evaluates by removing each feature one by one and continues to eliminate features until it selects the best feature subset.
3. Genetic Search applies genetic algorithm (GA) to search feature space. Each state is defined by chromosome that actually represents a feature subset. With this representation, implementation of GA for feature selection becomes quite simple. However, the evaluation of fitness function, i.e. its classification accuracy, can be expensive.
4. Simulated annealing (SA), in contrast to GA which maintains the population of chromosomes (each chromosome represents a feature subset), considers only one solution. It implements a stochastic search as there is a chance that some deterioration in solution is accepted—this allows a more effective exploration of the search space.

Forward elimination and backward elimination terminate when adding or deleting further features do not affect classification accuracy. However, these greedy search strategies do not ensure the best feature subset. GA and SA can be more sophisticated approaches and can be used to explore search space in a better way.

2.3.4 Feature Selection in Unsupervised Learning

Feature selection in unsupervised learning can, however, be challenging because the success criterion is not clearly defined. Various unsupervised feature selection techniques have been proposed in literature, e.g. [45–47], however, in the next section, we will discuss only a few representative techniques. Feature selection in unsupervised learning has been classified in the same way as in supervised learning. Two categories are unsupervised filters and unsupervised wrappers as discussed below.

2.3.4.1 Unsupervised Filters

The main characteristics of filter-based approaches are that features are selected on the basis of some rank or score which remains independent of the classification or clustering process. Laplacian score (LS) is one of the examples of this strategy, which can be used for DR when motivation is that the locality is preserved. The LS uses this idea for unsupervised feature selection [48]. LS selects features by preserving the distance between objects both in input and reduced output space. This criterion presumes all the features are relevant; the only thing is that they may just be redundant.

LS is calculated using a graph G that realizes nearest neighbour relationships between input data points. A square matrix S is used to represent G where $S_{ij} = 0$ unless x_i and x_j are neighbours, in which case:

$$s_{ij} = e^{-\frac{x_i - x_j^2}{t}}$$
(2.5)

Here, 't' is a bandwidth parameter. L = D − S represents Laplacian of the graph and D = degree of diagonal matrix as given below

$$D_{ij} = \sum_j S_{ij}, D_{ij,\, i \neq j} = 0$$
(2.6)

LS can be calculated using the following calculations:

$$\tilde{m}_i = m_i \frac{m_i^T}{1^T D 1} 1 \tag{2.7}$$

$$LS_i = \frac{\tilde{m}_i^T L \tilde{m}_i}{\tilde{m}_i^T D \tilde{m}_i} \tag{2.8}$$

where m_i is the vector of values for the ith feature and 1 is a vector or 1 s of length n.

All the features can be scored on this criterion, i.e. how efficiently they preserve locality. This idea can be appropriate for domains where locality preservation is an effective motivation [48], e.g. image analysis. However, it may not be a sensible motivation in case of irrelevant features, e.g. in analysis of gene expression data or text classification.

2.3.4.2   Unsupervised Wrappers

Wrapper-based techniques use classification or clustering process as part of feature selection to evaluate feature subsets. One such technique is proposed in [49]. Authors have used the notion of a category unit (CU) [40] to present unsupervised wrapper-like feature subset selection algorithm. CU was used as evaluation function to guide the process of creating concepts and can be defined as follows:

$$CU(C,F) = \frac{1}{k} \sum_{q \in C} \left[\sum_{f_i \in F} \sum_{j=1}^{r_i} P(f_{ij}|C_l)^2 - \sum_{f_i \in F} \sum_{j=1}^{r_i} P(f_{ij})^2 \right] \tag{2.9}$$

Here,
$C = \{C_1, \ldots \ldots C_l, \ldots \ldots C_k\}$ is the set of clusters
$F = \{F_1, \ldots \ldots F_i, \ldots \ldots F_p\}$ is the set of features.

CU calculates the difference between the conditional probability of a feature i having value j in cluster l and its prior probability. The innermost sum is over r feature values, the middle sum is over p features and the outer sum is over k clusters. CU is used as a key concept to score the quality of clustering in a wrapper-like search.

2.3.4.3   The Embedded Approach

Embedded approach is the final category of feature selection techniques. Just like construction of decision tree, feature selection in embedded approaches is an integral part of the classification algorithm. There are various techniques that perform feature selection in this category, e.g. [51–53]. Here, we will discuss ESFS

[54], an embedded feature selection approach which incrementally adds most relevant features. Process comprises of four steps:

Step 1: Calculate belief masses of the single features.
Step 2: Evaluate single features and select initial set of potential features.
Step 3: Combine features to generate feature subsets.
Step 4: Evaluate Stopping criterion and select best feature subset.

The stop criterion of ESFS occurs when the best classification rate begins to decrease while increasing the size of the feature subsets. Inspired from wrapper SFS, the algorithm incrementally selects features. It makes use of the term 'belief mass' for feature processing introduced from the evidence theory, which allows to merge feature information in an embedded way.

2.4   Correlation-Based Feature Selection

Feature selection techniques that we have discussed till now, majority of them consider single attributes. However, and important category of the feature selection techniques is the one that considers more than one attributes (i.e. complete attribute subset) at a time while interpreting the relation between attributes themselves along with the relation between attributes and decision class. We call this category as correlation-based feature selection (CFS). CFS is based on the following assumption [55].

Good feature subsets contain features highly correlated with the class, yet uncorrelated with each other.

Equation for CFS is [55]

$$r_{zc} = \frac{k_{\overline{r_{zi}}}}{\sqrt{k + k(k-1)_{\overline{r_{ii}}}}} \tag{2.10}$$

Here,

r_{zc} = Correlation between the summed components and the outside variable,
k = Number of components,
$\overline{r_{zi}}$ rzi is the average of the correlations between the components and the outside variable,
$\overline{r_{ii}}$ = average intercorrelation between components
Following are some of the heuristics that may derive a CFS algorithm:

- Higher correlation between the components and the outside variable results in higher correlation between composite and outside variable.
- Lower intercorrelations among the components results in higher correlation between the composite and the outside variable.

- Increasing number of components in the composite increases the correlation between the composite and the outside variable.

2.4.1 Correlation-Based Measures

Mainly there are two approaches to measure the correlation between two random variables, the one is based on linear correlation and other is based on Information Theory [56]. Under the first approach, linear correlation coefficient is one of the well-known measures. For two variables X and Y, the linear correlation coefficient is

$$r = \frac{\sum_i (x_i - \bar{x}_i)(y_i - \bar{y}_i)}{\sqrt{\sum_i (x_i - \bar{x}_i)^2} \sqrt{\sum_i (y_i - \bar{y}_i)^2}} \tag{2.11}$$

Here,

$\bar{x}_i$ = mean of X
$\bar{y}_i$ = mean of Y

r can take the value between -1 and 1. If X and Y are fully correlated r will either be 1 or -1 and 0 if X and Y are independent.

Linear correlations help identify redundant features and those having zero linear correlation, however, linear correlation measure cannot help in case of non-linear correlations.

The other approach based on Information Theory uses the concept of information entropy. Entropy defines the measure of uncertainty of a random variable. Mathematically, entropy X of a random variable is

$$H(X) = -\sum P(x_i) log_2(P(x_i)) \tag{2.12}$$

Entropy of X given Y is

$$H(X|y) = -\sum_j P(x_i) \sum_i P(x_i|y_j) log_2\left(P(x_i|y_j)\right) \tag{2.13}$$

Here,
$P(x_i)$ = prior probabilities for all values of X,
$P(x_i|y_i)$ = posterior probabilities of X given the values of Y.

The amount by which the entropy of X decrease after variable Y is called Information Gain. Mathematically.

$$IG(X|Y) = H(X) - H(X|Y) \tag{2.14}$$

Now, we present a few of the approaches based on correlation-based feature selection.

2.4.1.1 Correlation-Based Filter Approach (FCBF)

FCBF [56] used Symmetrical Uncertainty (SU) to evaluate the goodness of a feature subset. Algorithm is based on the following definitions and heuristics [58]:

Definition 1 (*Predominant Correlation*) The correlation between a feature $F_i(F_i \in S)$ and the class C is predominant iff $SU_{i,c} \geq \partial(F_i \in S)$ and $\forall F_j \in S'(j \neq i)$, there exists no F_j such that $SU_{j;i} \geq SU_{i;c}$.

Definition 2 (*Predominant Feature*) A feature is predominant to the class, if its correlation to the class is predominant or can become predominant after removing its redundant peers.

Heuristic 1 (if $S_{Pi}^+ = \phi$) Treat F_i as a predominant feature, remove all features in S_{Pi}, and skip identifying redundant peers for them.

```
input: S(F₁, F₂, ..., F_N, C)      // a training data set
δ                                   // a predefined threshold
output: S_best                      // an optimal subset

(1)  begin
(2)  for i = 1 to N do begin
(3)     calculate SU_{i,c} for F_i;
(4)     if(SU_{i,c} ≥ δ)
(5)        append F_i to S'_list;
(6)  end;
(7)  order S'_list in descending SU_{i,c} value;
(8)  F_p = getFirstElement(S'_list);
(9)  do begin
(10)       F_q = getNextElement(S'_list, F_p);
(11)       if(F_q <> NULL)
(12)       do begin
(13)          F'_q = F_q;
(14)          if(SU_{p,q} ≥ SU_{q,c})
(15)          remove F_q from S'_list;
(16)          F_q = getNextElement(S'_list, F'_p);
(17)          else F_q = getNextElement(S'_list, F_p);
(18)       end until (F_q == NULL);
(19)       F_p = getNextElement(S'_list, F_p);
(20)   end until (F_p == NULL);
(21)   S_best = S'_list;
(22)  end;
```

Fig. 2.8 Pseudocode of the FCBF algorithm

Heuristic 2 (if $S_{Pi}^{+} \neq \phi$) Process all features in S_{Pi}^{+} before making a decision on F_i. If none of them becomes predominant, follow Heuristic 1; otherwise only remove F_i and decide whether or not to remove features in S_{Pi}^{-} based on other features in S'.

Heuristic 3 (starting point) The feature with the largest $Su_{i,c}$ value is always a predominant feature and can be a starting point to remove other features.

Figure 2.8 shows the pseudocode of the FCBF algorithm

Algorithm comprises of two parts. In the first part, it arranges relevant features in S'list based on predefined threshold by calculating the SU value of each feature, which are then arranged in decreasing order of their SU value. In the second part, it removes the redundant features from S'list by keeping only the predominant ones. Algorithm continues until there are no more features to remove. Complexity of the first part of algorithm is N whereas that of the second part is O(N logN), Since the calculation of SU for a pair of features is linear in term of the number of instances M in a dataset, the overall complexity of FCBF is O(MN logN).

2.4.2   Efficient Feature Selection Based on Correlation Measure (ECMBF)

The authors in [57] have presented an efficient correlation (between continuous and discrete measure) based feature selection algorithm. Algorithm uses Markov Blanket to determine feature redundancy. Correlation between continuous and discrete measures (CMCD) was used for feature selection.

Algorithm takes dataset, relevance threshold α and the redundancy threshold β as input. Threshold value of α is used to determine feature relevance. If the correlation between feature and class is lesser than α, it is considered to be less relevant or irrelevant thus should be removed. If correlation between two random features is greater than β, it means one of them is redundant and thus should be removed. In that case, the feature with lower correlation value is removed.

As the threshold values of threshold help select the optimal feature subset, so these values should be carefully selected. Regarding the time complexity of ECMBF, it comprises of two aspects. First, the time complexity of calculating correlation between any two features which measures to be $O(m*(m - 1))$. Worst-case time complexity for all the features will be $O(n * m2)$. Second to select the relevant features, time complexity will be $O(m * \log_2 m)$. The overall time complexity will be: Thus, the overall time complexity of ECMBF will be $O(n * m2)$. Figure 2.9 shows the pseudocode of the ECMBF algorithm.

Fig. 2.9 Pseudocode of
ECMBF algorithm

```
input: F(X₁, X₂, L, Xₘ, C)       // a training data set
α, β                   // relevance threshold and redundant threshold
output: Subset                   // an optimal subset
  (1) begin
  (2) for i = 1 to m do begin
  (3) calculate sim(Xᵢ, C) and sim(Xᵢ, Xⱼ);
  (4) order features in ascending sim(Xᵢ, C)value, and append
      Xᵢto S_rank-list;
  (5) find the greatest decent point as the relevant threshold α
      from S_rank-list
  (6) if (sim(Xᵢ, C) ≤ α)
  (7) remove Xᵢ from S_rank-list, the new subset denote as S'_rank-list
      = (X₁, X₂, ..., Xₖ), for k < m;
  (8) end;
  (9) Xₚ = getNextElement(S'_rank-list)
 (10)     do begin
 (11)         Xₑ = getNextElement(S'_rank-list- Xₚ);
 (12)         if(Xₑ<> NULL)
 (13)         do begin
 (14)         X'ₑ = Xₑ;
 (15)         if(sim(Xₚ, Xₑ) > β)
 (16)         remove Xₑ from S'_rank-list;
 (17)         Xₑ = getNextElement(S'_rank-list- X'ₑ);
 (18)         else Xₑ = getNextElement(S_rank-list - Xₑ);
 (19)         end until (Xₑ == NULL);
 (20)         Xₚ = getNextElement(S'_rank-list -Xₚ);
 (21)         end until (Xₚ == NULL);
 (22)         Subset = S'_rank-list;
 (23)         end;
```

2.5   Mutual Information-Based Feature Selection

Mutual information-based feature selection is another commonly used feature
selection mechanism. Various algorithms in literature have been proposed using
this measure. While correlation measures the linear or monotonic relationship
between two variables. MI is more generic form to measure the decrease in
uncertainty in variable X after observing variable Y. Mathematically:

$$I(X;Y) = \sum_{x,y} P(x,y) log \frac{P(x,y)}{P(x)P(y)} \qquad (2.15)$$

Here,

p(x, y) = joint probability distribution function of X and Y,

p(x) and p(y) are the marginal probability distribution functions of X and Y.

Mutual Information can also be defined in terms of entropy. Here are some
mathematical definitions of MI in terms of MI.

$$I(X;Y) = H(X) - H(X|Y) :$$
$$I(X;Y) = H(Y) - H(Y|X) :$$
$$I(X;Y) = H(X) + H(Y) - H(X,Y) :$$
$$I(X;Y) = H(X,Y) - H(X|Y) - H(Y|X) :$$

Here, H(X) and H(Y) are marginal entropies. Here, we will present few MI-based feature selection algorithms.

2.5.1   A Mutual Information-Based Feature Selection Method (MIFS-ND)

MIFS-ND [58] considers both Feature-Feature MI and Feature-Class MI. Given a dataset, MIFS-NS, first computes feature-feature mutual information to select the feature that has highest MI value, removes it from original set and puts in selected feature list. From the remaining feature subset, it then calculates feature-class mutual information and average feature-feature mutual information measure for each of the selected feature. Till this point, each selected feature has average feature-feature mutual information measure and each non-selected feature has feature-class mutual information. Now from these values, it selects feature with highest feature-class mutual information and minimum feature-feature mutual information

Algorithm then uses two terms:

- Domination count: domination count represents the number of features that a feature dominates for feature-class mutual information.
- Dominated count: dominated count represents the total number of features that a feature dominates for feature-feature mutual information.

Algorithm then selects the features that have maximum difference of domination count and dominated count. The reason behind is to select feature that is either strongly relevant or weakly redundant. Complexity of the MIFS-NS algorithm depends on dimensionality of the input dataset. For a dataset with dimensions 'd' the computational complexity of algorithm to select a subset of relevance features is $O(d^2)$. Figure 2.10 shows the pseudocode of the proposed algorithm.

Algorithm uses two major modules, Compute_FFMI which computes feature-feature mutual information and Compute_FCMI which computes feature-class mutual information. So, by using these two modules, algorithm selects the features that are highly relevant and non-redundant.

```
input: d, the number of features; dataset D;
F = {f₁, f₂, ..., f_d}, the set of features
output: F', an optimal subset of features
Steps:
    (1) for i = 1 to d,do
    (2) compute MI(fᵢ, C)
    (3) end
    (4) select the features fᵢ with maximum MI(fᵢ, C)
    (5) F' = F'U { fᵢ}
    (6) F = F - { fᵢ}
    (7) count = 1;
    (8) while count <= k do
    (9) for each feature fⱼ Ɛ F, do
   (10)        FFMI = 0;
   (11)        for each feature fⱼ Ɛ F', do
   (12)            FFMI = FFMI + compute_FFMI(fᵢ, fⱼ)
   (13)        end
   (14)        AFFMI = Average FFMI for feature fⱼ.
   (15)        FCMI = compute_FCMI(fᵢ, C)
   (16)        end
   (17)        select the next feature fⱼ that has maximum AFFMI but
               minimum FCMI
   (18)        F' = F'U { fᵢ}
   (19)        F = F - { fⱼ}
   (20)        i = j
   (21)        count = count + 1;
   (22)        end
   (23)        Return features set F'
```

Fig. 2.10 Pseudocode of MIFS-ND algorithm

1. (Initialisation) Set $F \leftarrow$ "initial set of n features"; $S \leftarrow$ "empty set."
2. (Computation of the MI with the output class) For $\forall f_i \, \mathcal{E} \, F$ compute $I(C: f_i)$.
3. (Choice of the first feature) Find a feature f_i that maximises $I(C: f_i)$; set $F \leftarrow F\backslash\{ f_i\}$; set $S \leftarrow \{ f_i\}$.
4. (Greedy selection) Repeat until $|S| = k$; (Selection of the next feature) Choose the feature
$$f_i = argmax_{f_i \in F-S}\left(min_{f_i \in S}(I(f_i. f_s: C))\right); \text{ set } F \leftarrow F\backslash\{ f_i\}; \text{ set } S \leftarrow S \, U \, \{f_i\}.$$
5. (Output) Output the set S with selected features.

Fig. 2.11 Forward greedy search

2.5.2 Multi-objective Artificial Bee Colony (MOABC) Approach

In [59], the authors propose a new method for feature selection based on joint mutual information maximization (jMIM) and normalized joint mutual information maximization (NjMIM). jMIM approach applies joint mutual information and maximum of minimum approach to choose the most relevant features. The intention of the proposed approach is to overcome the problem of overestimating significance of some of the features that occurs during cumulative summation approximation. jMIM uses the following forward greedy search strategy (shown in Fig. 2.11) to find a feature subset of size K

The second approach proposed i.e. NjMIM intended to study the effect of using normalized MI instead of MI. It uses the similar goal function as that used in jMIM except that symmetrical relevance is used instead of MI. The same forward greedy search strategy was used to search feature subset for the given dataset.

2.6   Summary

In this chapter, we have provided a detailed description of various feature selection techniques proposed in literature from time to time. A complete taxonomy of feature selection was presented and feature selection algorithms according to each category in this taxonomy were explained. Efforts were made to provide the pseudocode as well along with the detailed description. Both transformation-based and selection-based techniques were discussed. This was the end of the first part in which we tried to provide a strong foundation of feature selection from basic concepts to its applications in real life. In the next part, we will start with Rough Set Theory.

References

1. Villars RL, Olofson CW, Eastwood M (2011) Big data: what it is and why you should care. IDC, White Paper, p 14
2. Asuncion A, Newman D (2007) UCI machine learning repository
3. Bellman R (1956) Dynamic programming and lagrange multipliers. Proc Natl Acad Sci 42 (10):767–769
4. Yan J et al (2006) Effective and efficient dimensionality reduction for large-scale and streaming data preprocessing. IEEE Trans Knowl Data Eng 18(3):320–333
5. Han Y et al Semisupervised feature selection via spline regression for video semantic recognition. IEEE Trans Neural Netw Learn Syst 26(2):252–264 (2015)
6. Boutsidis C et al. (2015) Randomized dimensionality reduction for k-means clustering. IEEE Trans Inf Theory 61(2):1045–1062
7. Cohen, MB et al (2015) Dimensionality reduction for k-means clustering and low rank approximation. In: Proceedings of the forty-seventh annual ACM on symposium on theory of computing. ACM
8. Bourgain J, Dirksen S, Nelson J (2015) Toward a unified theory of sparse dimensionality reduction in euclidean space. Geom Funct Anal 25(4):1009–1088
9. Radenović F, Jégou H, Chum O (2015) Multiple measurements and joint dimensionality reduction for large scale image search with short vectors. In: Proceedings of the 5th ACM on international conference on multimedia retrieval. ACM

10. Azar AT, Hassanien AE (2015) Dimensionality reduction of medical big data using neural-fuzzy classifier. Soft Comput 19(4):1115–1127
11. Pawlak Z (1991) Rough sets: theoretical aspects about data. Springer, Cham
12. Qian Y et al (2015) Fuzzy-rough feature selection accelerator. Fuzzy Sets Syst 258:61–78
13. Tan A et al (2015) Matrix-based set approximations and reductions in covering decision information systems. Int J Approx Reason 59:68–80
14. Al Daoud E (2015) An efficient algorithm for finding a fuzzy rough set reduct using an improved harmony search. Int J Modern Educ Comput Sci 7(2):16
15. Candès EJ et al (2011) Robust principal component analysis? J ACM (JACM) 58(3), 11
16. Kao Y-H, Benjamin Van R (2013) Learning a factor model via regularized PCA. Mach Learn 91(3):279–303
17. Varshney KR, Willsky AS (2011) Linear dimensionality reduction for margin-based classification: high-dimensional data and sensor networks. IEEE Trans Signal Process 59 (6):2496–2512
18. Van Der Maaten L, Postma E, Van den Herik J (2009) Dimensionality reduction: a comparative. J Mach Learn Res 10:66–71
19. Jensen R (2005) Combining rough and fuzzy sets for feature selection. Dissertation University of Edinburgh
20. Cunningham P (2008) Dimension reduction. In: Machine learning techniques for multimedia, pp 91–112. Springer, Berlin
21. Friedman JH, Stuetzle W (1981) Projection pursuit regression. J Am Stat Assoc 76(376):817–823
22. Borg I, Patrick JF (2005) Modern multidimensional scaling: theory and applications. Springer Science & Business Media, New York
23. Dalgaard, Peter. Introductory statistics with R. Springer Science & Business Media, 2008
24. Gisbrecht A, Schulz A, Hammer B (2015) Parametric nonlinear dimensionality reduction using kernel t-SNE. Neurocomputing 147:71–82
25. Gottlieb L-A, Krauthgamer R (2015) A nonlinear approach to dimension reduction. Discrete Comput Geometry 54(2):291–315
26. Gisbrecht A, Hammer B (2015) Data visualization by nonlinear dimensionality reduction. Wiley Interdiscip Rev Data Mining Knowl Discov 5(2):51–73
27. Zeng X, Luo S (2008) Generalized locally linear embedding based on local reconstruction similarity. In: 2008 Fifth international conference on fuzzy systems and knowledge discovery, FSKD'08, vol 5. IEEE
28. Saul LK et al (2006) Spectral methods for dimensionality reduction. Semisupervised Learn 293–308
29. Liu R et al (2008) Semi-supervised learning by locally linear embedding in kernel space. In: 2008 19th international conference on pattern recognition, ICPR 2008. IEEE
30. Gerber S, Tasdizen T, Whitaker R (2007) Robust non-linear dimensionality reduction using successive 1-dimensional Laplacian eigenmaps. In: Proceedings of the 24th international conference on machine learning. ACM
31. Teng L et al (2005) Dimension reduction of microarray data based on local tangent space alignment. In: 2005 fourth IEEE conference on cognitive informatics (ICCI 2005). IEEE
32. Dimensionality reduction methods for molecular motion. http://archive.cnx.org/contents/02ff5dd2-fe30-4bf5-8e2a-83b5c3dc0333@10/dimensionality-reduction-methods-for-molecular-motion. Accessed 30 March 2017
33. Balasubramanian M, Schwartz EL (2002) The isomap algorithm and topological stability. Science 295(5552):7–7
34. Faraway JJ (2005) Extending the linear model with r (texts in statistical science)
35. Jensen R, Shen Q (2008) Computational intelligence and feature selection: rough and fuzzy approaches, vol 8. Wiley
36. Cunningham P (2008) Dimension reduction: machine learning techniques for multimedia, pp 91–112. Springer, Berlin

37. Tang B, Kay S, He H (2016) Toward optimal feature selection in naive Bayes for text categorization. IEEE Trans Knowl Data Eng 28(9):2508–2521

38. Jiang F, Sui Y, Zhou L (2015) A relative decision entropy-based feature selection approach. Pattern Recogn 48(7):2151–2163

39. Singh DA et al (2016) Feature selection using rough set for improving the performance of the supervised learner. Int J Adv Sci Technol 87:1–8

40. Xu J et al (2013) L_1 graph based on sparse coding for feature selection. In: International symposium on neural networks. Springer, Berlin

41. Almuallim H, Dietterich TG (1991) Learning with many irrelevant features. In: AAAI, vol 91

42. Kira K, Rendell LA (1992) The feature selection problem: traditional methods and a new algorithm. In: AAAI, vol 2

43. Raman B, Ioerger TR (2002) Instance-based filter for feature selection. J Mach Learn Res 1 (3):1–23

44. Liu H, Motoda H (2007) (eds) Computational methods of feature selection. CRC Press

45. Du L, Shen Y-D (2015) Unsupervised feature selection with adaptive structure learning. In: 2015 Proceedings of the 21th ACM SIGKDD international conference on knowledge discovery and data mining. ACM

46. Li J et al (2015) Unsupervised streaming feature selection in social media. In: Proceedings of the 24th ACM international on conference on information and knowledge management. ACM

47. Singh DA, Balamurugan SA, Leavline EJ (2015) An unsupervised feature selection algorithm with feature ranking for maximizing performance of the classifiers. Int J Autom Comput 12 (5):511–517

48. He X, Cai D, Niyogi P (2005) Laplacian score for feature selection. In: NIPS, vol 186

49. Devaney M, Ram A (1997) Efficient feature selection in conceptual clustering. In: ICML, vol 97

50. Gluck M (1985) Information, uncertainty and the utility of categories. In: Proceedings of the seventh annual conference on cognitive science society. Lawrence Erlbaum

51. Yang J, Hua X, Jia P (2013) Effective search for genetic-based machine learning systems via estimation of distribution algorithms and embedded feature reduction techniques. Neurocomputing 113:105–121

52. Imani MB, Keyvanpour MR, Azmi R (2013) A novel embedded feature selection method: a comparative study in the application of text categorization. Appl Artif Intell 27(5):408–427

53. Viola M et al (2015) A generalized eigenvalues classifier with embedded feature selection. Optim Lett 1–13

54. Xiao Z et al (2008) ESFS: a new embedded feature selection method based on SFS. Rapports de recherché

55. Hall MA (2000) Correlation-based feature selection of discrete and numeric class machine learning

56. Yu L, Liu H (2003) Feature selection for high-dimensional data: a fast correlation-based filter solution. In: ICML, vol 3

57. Jiang S-y, Wang L-x (2016) Efficient feature selection based on correlation measure between continuous and discrete features. Inf Process Lett 116(2):203–215

58. Hoque N, Bhattacharyya DK, Kalita JK (2014) MIFS-ND: a mutual information-based feature selection method. Expert Syst Appl 41(14):6371–6385

59. Hancer E et al (2015) A multi-objective artificial bee colony approach to feature selection using fuzzy mutual information. In: 2015 IEEE congress on evolutionary computation (CEC). IEEE

Chapter 3
Rough Set Theory

This chapter discusses the basic preliminaries of rough set theory (RST). Since its inception, RST has been a prominent tool for data analysis due to its analysis friendly nature. RST provides a range of data structures, e.g. Information Systems, Decision Systems and Approximations to represent the real-world data. Furthermore, it provides various methods to help analyse this data. This chapter discusses the basic concepts of RST with an example to set a strong foundation of RST to be used as feature selection.

3.1 Classical Set Theory

Classical set theory is a branch of mathematics that deals with the collection of objects called sets. These objects are called members of that set. Rough Set Theory may be called as the extension of classical set theory. The main problem with classical set theory is that it is concrete in nature and hence fails to model the vagueness of the universe, this is where Rough Set theory plays its role. Before discussing the Rough Set Theory, here we will discuss some basics of Classical Set Theory.

3.1.1 Sets

A set is a well-defined collection of distinct objects. 'Well-defined' here means that we should clearly be able to distinguish either an object belongs to a set or not. For example, a set of even numbers between 10 and 100 is a set because every number can precisely be concluded as belonging to this set or not. On the other hand, 'a set of intelligent people' is not a set as there is no specific rule to specify either a person

© Springer Nature Singapore Pte Ltd. 2019
M. S. Raza and U. Qamar, *Understanding and Using Rough Set Based Feature Selection: Concepts, Techniques and Applications*,
https://doi.org/10.1007/978-981-32-9166-9_3

is intelligent or not'. 'Distinct' means that an object will appear only once and cannot be repeated in a set.

Some examples of sets are as follows:

(i)   Set of supports goods.
(ii) Set of animals that lay eggs.
(iii) Set of schools in New York.
(iv) Set of students who scored greater than 85% score in Chap. 2

If you note, any object can be clearly defined as belonging to these sets or not. Similarly, the members of these sets will be distinct and will not be repeated.

3.1.2   SubSets

A set 'A' will be subset of another set 'B' if every member of 'A' is present in 'B'. Mathematically, $A \subseteq B$, here the symbol '$\subseteq$' represents subset. We will provide details of all of the symbols used in the upcoming section.

Example:

Consider the following three sets:

A = {2, 4, 6, 8, 10}, B = {1, 2, 4} and C = {6, 8, 10}

Here,

(i) $C \subseteq A$, i.e. 'C' is subset of 'A' as all members of 'C' are present in 'A'.
(ii) $B \nsubseteq A$, i.e. 'B' is not subset of 'A' as all members of 'B' are not present in 'A'.
(iii) $A \subseteq A$, $B \subseteq B$ and $C \subseteq C$, i.e. a set is its own subset as all of its members are present in it.

3.1.3   Power Sets

For a set A, there exist a set whose elements are all subsets of 'A'. Here, the elements of power subsets will not be individual objects but subsets.

Example:

Consider the Set A = {1, 2, 3}

P(A) = {{ }{1}{2}{3}{1, 2}{1, 3}{2, 3}{1, 2, 3}}

P(A) comprises of all of subsets that can be formed by different arrangements of all elements of 'A'.

3.1.4 Operators

Different operators exist for manipulation of Sets. Here we will discuss some basic operators

3.1.4.1 Intersection

Sometimes we need the elements common in all sets. Intersection operator is intended for this purpose. By definition, Intersection of two sets 'A' and 'B' is another set 'C' that contains all the elements common in 'A' and 'B'. Intersection is denoted by the symbol ' $\cap$ '. Mathematically, $A \cap C = \{x | (x \in A) \text{ and } (x \in B)\}$.

Example:

A = {1, 2, 3, 4, 5, 6}
B = {4, 5, 6, 7, 8, 9}
C = {0, 1, 2, 3}

(1) A $\cap$ B = {4, 5, 6}
(2) A $\cap$ C = {1, 2, 3}
(3) A $\cap$ C = {1, 2, 3}

Note: The symbol ' ϕ ' represents empty set.

3.1.4.2 Union

Union of two sets 'A' and 'B' is another set 'C' that contains all the elements of 'A' and 'B' such that if an element appears in both sets, it is taken once. Union is denoted by symbol ' $\cup$ '. Mathematically: $A \cup C = \{x | (x \in A) \text{ or } (x \in B)\}$, i.e. to belong to union of sets 'A' and 'C', an element should be member of either of 'A' or 'B'.

A = {1, 2, 3, 4, 5, 6}
B = {4, 5, 6, 7, 8, 9}
C = {0, 1, 2, 3}

(1) $A \cup B = \{1,2,3,4,5,6,7,8,9\}$
(2) $A \cup C = \{0,1,2,3,4,5,6\}$
(3) $B \cup C = \{0,1,2,3,4,5,6,7,8,9\}$

Note that in '4,5,6' appear both in 'A' and 'B', however, in $A \cup B$ they are written once.

3.1.4.3 Complement

The complement of 'A' in 'B' is a set 'C' that contains the elements present in 'A' but not in 'B'. It is denoted by 'A–B' or 'A\B'. Normally, we consider all the sets to be subsets of universal set 'U'. So, U—A becomes the absolute complement of 'A'.

A = {2, 3, 4}
B = {0, 1, 2, 3}

(1) $A - B = \{4\}$
(2) $B - A = \{0, 1\}$

3.1.4.4 Cardinality

Cardinality of a set is the total number of objects present in that set, e.g. symbolically cardinality of set 'A' will be represented as '|A|'. Consider the following set:

A = {2, 4, 6, 8, 10}
A = {1, 3, 5}
|A| = 5
|B| = 3

3.1.5 Mathematical Symbols for Set Theory

Here is a list of symbols used in set theory taken from [1]. Memorizing these symbols is helpful as they are also used in Rough Set Theory.

3.2 Knowledge Representation and Vagueness

Knowledge is the ability to classify objects. An object here simply refers to its classical definition that is 'an object is abstraction or realization of real world entities that show some properties'. The properties are represented in the form of attributes'. We have already discussed attributes in detail in the first chapter.

These collection of objects is called universe, they form sets called universe of discourse. Examples may be a set of students, set of medicines and set of supports goods, etc. Classification of objects in simple words refers to finding subsets of the universe having objects that have common values regarding the concept under consideration, e.g. from a set of fruits having different fruits, classifying the objects that belong same season or have the same colour. The clearest and unambiguous

Fig. 3.1 Classification of objects given in Table 3.2

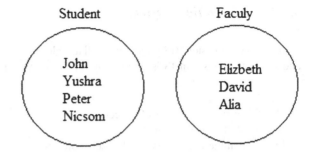

classification is when an object belongs to a single classification under the single concept. For example, consider Table 3.2 given below:

Now if we classify the above set of universe with respect to 'Status' we will have two sharp classifications as follows (Fig. 3.1):

But what if David is also a student taking some professional course. Will it belong to 'Student' classification or 'Faculty'. This example also shows the vagueness we have in our real world. Unfortunately, classical set theory fails to represent this vagueness. It is concrete in nature, here an object can either belong to a set or not, representing information in this way, much of the information is lost. For example, consider the set:

A = {Students with CGPA $\geq$ 3.5} = {John, Peter, Elizbeth}

By the means of classical set theory, both 'John' and 'Peter' are equally competent (with respect to grades), however, the fact that 'Peter' has more grades than 'John' cannot be formulated here. Some extensions of classical set theory have proposed including Rough Set Theory, Fuzzy Set Theory and its hybridization with Rough Set Theory, i.e. Fuzzy Rough Set theory. Complete description of Fuzzy Set Theory is out of scope of this book, however, in next chapter we will give some introduction of it to discuss Fuzzy Rough Set Theory.

3.3   Rough Set Theory (RST)

Proposed by ZidslawPawlak [2], RST has become a topic of great interest over the past ten years and has been successfully applied to many domains by researchers. As discussed above classical sets are concrete in nature, so fail to model the vagueness of real world. RST overcomes the problem by the concept of set approximations (discussed later). RST is a combination of data structures and tools/techniques that make is analysis friendly. Here, we will discuss some preliminaries of RST.

3.3.1   Information Systems

An information system (IS) is just like a flat table or view [2] comprising of objects
and their attributes. An IS is defined by a pair (U,A) [2] as given below

   IS = (U,A)

Here,

U = finite non empty set of objects

A = attributes of the objects

Every attribute $a \in A$ has a value set represented by V_a as shown below. Each
value set of an attribute contains all possible values of that attribute.

$$a : U \rightarrow V_a$$

Table 3.3 is an information system IS = (U,A) where
$U = \{x_1,\ x_2,\ x_3,\ x_4,\ x_5,\ x_6,\ x_7\}$
$A = \{Age,\ Incom\}$.

3.3.2   Decision Systems

Decision systems (DS) [2] are a special form of Information System having decision attribute also called the class of the object. Every object belongs to a specific
class. The value of the class depends on other attributes called conditional attributes. Formally:

$$\alpha = (U, C \cup \{D\})$$

where:

C = set of conditional attributes

D = Decision attribute (or class)

   Table 3.4 shows a decision system with the policy as decision attribute (or
class).

3.3.3   Indiscernibility

A decision system represents all knowledge about a model. This table may be
unnecessarily large in two ways: there may be identical or indiscernible objects
having more than one occurrence and there may be superfluous attributes. The
notion of equivalence is recalled first. A binary relation is called equivalence

relation if it is reflexive, i.e. an object is in relation with itself xRx, symmetric, i.e. if xRy, then yRx and transitive, i.e. if xRy and yRz then xRz. The equivalence class of an element consists of all objects such that xRy.

Let $A = (U, C \cup \{D\})$ be a decision system; indiscernibility defines an equivalence relation between objects in A. For any $c \in C$ in A, there exists an indiscernibility relation $IND_A(C)$:

$$IND_A(C) = \{ (O_1 = O_2) \in U^2 | \forall c \in Cc(O1) = c(O2) \} \tag{3.1}$$

$IND_A(C)$ (also denoted by $[x]_c$) is called a 'C-indiscernibility' relation. If two objects $(O_1, O_2) \in IND_A(C)$, then these objects are indiscernible or indistinguishable with respect to C. Considering Table 3.3, objects x_1, x_2 are indiscernible with respect to attribute 'Age'. Similarly, objects x_3 and x_5 are indiscernible with respect to attribute 'Income'. The subscript is normally omitted if we are sure about which information system is meant. In Table 3.3:

$IND(\{Age\}) = \{\{x_1, x_2\}, \{x_3, x_5\}, \{x_4, x_6, x_7\}\}$
$IND(\{Income\}) = \{\{x_1, x_2\}, \{x_3, x_5\}, \{x_4, x_6, x_7\}\}$

3.3.4 Approximations

Most of the sets cannot be identified unambiguously, so we use approximation. For an information system where $B \subseteq A$, we can approximate the decision class X by using the information contained in B. The lower and upper approximations are defined as follows [2]:

$$\underline{B}X = \{x | [x]_B \subseteq X\} \tag{3.2}$$

$$\bar{B}X = \{x | [x]_B \cap X \neq \emptyset\} \tag{3.3}$$

Lower approximation defines the objects that are definitely a member of X with respect to information in 'B'. Upper approximation, on the other hand, contains objects that with respect to 'B' can possibly be members of 'X'. The boundary region defines the difference between lower and upper approximation.

$$BN_B(X) = \bar{B}X - \underline{B}X \tag{3.4}$$

Using the decision system shown in Table 3.1(b), the situation can be sketched as Fig. 3.2. The lower boundary of 'Policy' defines all the equivalence classes that can surely belong to class Policy = Gold. Upper boundary defines classes that can possibly belong to Policy = Gold.

Table 3.1 Mathematical symbols and their meanings [1]

Symbols	Description	Symbol	Description
{ }	Set	$A \times B$	Cartesian product
$A \cap B$	Intersection	$|A|$	Cardinality
$A \cup B$	Union	$\#A$	Cardinality
$A \subseteq B$	Subset	$\aleph_0$	Aleph-null
$A \subset B$	Proper subset/strict subset	$\aleph_1$	Aleph-one
$A \not\subset B$	Not subset	$\varnothing$	Empty set
$A \supseteq B$	Superset	$\mathbb{U}$	Universal set
$A \supset B$	Proper superset/strict superset	$\mathbb{N}_0$	Natural numbers/whole numbers set (with zero)
$A \not\supset B$	Not superset	$\mathbb{N}_1$	Natural numbers/whole numbers set (without zero)
$2A$	Power set	$\mathbb{Z}$	Integer numbers set
$\mathcal{A}$	Power set	$\mathbb{Q}$	Rational numbers set
$A = B$	Equality	$\mathbb{R}$	Real numbers set
A^c	Complement	$\mathbb{C}$	Complex numbers set
$A \backslash B$	Relative complement	$a \in A$	Element of
$A\text{–}B$	Relative complement	$x \notin A$	Not element of
$A \Delta B$	Symmetric difference	(a, b)	Ordered pair
$A \ominus B$	Symmetric difference		

Table 3.2 Set of persons

Person	Name	Status
X_1	John	Student
X_2	Elizbeth	Faculty
X_3	David	Faculty
X_4	Yushra	Student
X_5	Peter	Student
X_6	Alia	Faculty
X_7	Nicson	Student

Table 3.3 Information system

Customer	Age	Income
X_1	35–40	30000–40000
X_2	35–40	30000–40000
X_3	40–45	50000–60000
X_4	25–35	20000–30000
X_5	40–45	50000–60000
X_6	25–35	20000–30000
X_7	25–35	20000–30000

Table 3.4 Decision system

Customer	Age	Income	Policy
X_1	35–40	30000–40000	Platinum
X_2	35–40	30000–40000	Platinum
X_3	40–45	50000–60000	Gold
X_4	25–35	20000–30000	Silver
X_5	40–45	50000–60000	Gold
X_6	25–35	20000–30000	Silver
X_7	25–35	20000–30000	Gold

Fig. 3.2 Approximation diagram

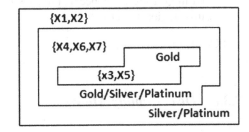

$$\underline{B}Policy = \{\{X_3, X_5\}\}$$

$$\bar{B}Policy = \{\{X_3, X_5\}, \{X_4, X_6, X_7\}\}$$

The boundary region is $\bar{B}Insurance - \underline{B}Insurance = \{X_4, X_6, X_7\}$. As it is non empty, so the set is rough set.

3.3.5 Positive Region

Lower approximation is also called positive region. Let P and Q be equivalence relations over U, then the positive region can be defined as

$$POS_P(Q) = \bigcup_{X \in U/D} \underline{P}(X) \tag{3.5}$$

where P is the set of conditional attributes and Q is the Decision class. The positive region is the union of all equivalence classes in $[X]_P$ that are subset of (or are contained by) target set.

Considering Table 3.1(b), we calculate the positive region for set Policy = 'Gold' as follows:

First, we will calculate $[X]_P$. Here,

$P_1 = \{x_1, x_2\}$
$P_2 = \{x_3, x_5\}$
$P_3 = \{x_4, x_6, x_7\}$

Now, we calculate $[X]_Q$ where Q implies the concept 'Insurance = Gold'. Here,

$Q = \{x_3, x_5, x_7\}$

It means we cannot distinguish between x_3, x_5 and x_7 with respect to information contained in Q. Here for concept 'Policy = Gold', only P_2 class belongs to Q. So, the positive region for Q will be

$POS_P(Q) = \{x_3, x_5\}$.

3.3.6 Discernibility Matrix

A discernibility matrix M of elements m_{ij} such that each element defines the set of attributes where the object $x_i, x_j \in U$ differ to each other. For an information system $A = (U, C \cup \{D\})$, the discernibility matrix is a n x n matrix with elements m_{ij} as given below.

$$x_{ij} = \{a \in A | a(xi) \neq a(xj)\} \text{ where } i, j = 1, 2, 3, \ldots, n \qquad (3.6)$$

Discernibility matrix is also a tool to find 'Reducts'. A Reduct is a subset of features that provide the same partition of universe as obtained by entire set of features. We will have an in-depth discussion on Reducts in upcoming topics. We will now explain discernibility matrix with an example, consider the following decision system given in Table 3.5.

The discernibility matrix for this decision system is given below in Table 3.6.

Consider the entry m31 (just leave the leftmost column and topmost row as these are for heading purpose only), i.e. the entry at intersection of object x3 and x1, the

Table 3.5 A sample decision system

	T	I	P	L	d
x1	1	1	1	2	1
x2	1	0	1	0	0
x3	2	0	1	1	0
x4	1	2	1	0	1
x5	1	1	1	0	0
x6	1	2	1	2	1
x7	1	2	0	1	1
x8	2	0	0	2	0

Table 3.6 Discernibility matrix

	X1	X2	X3	X4	X5	X6	X7	X8
x1	φ							
x2	(I∨L)	Φ						
x3	(T∨I∨L)	(T∨L)	Φ					
x4	(T∨I∨L)	(T∨I)	(T∨I∨L)	φ				
x5	(T∨L)	(T∨I)	(T∨I∨L)	(I)	φ			
x6	(T∨I)	(T∨I∨L)	(T∨I∨L)	(L)	(I∨L)	φ		
x7	(I∨P∨L)	(I∨P∨L)	(T∨I∨P)	(T∨P∨L)	(T∨I∨P∨L)	(T∨P∨L)	φ	
x8	(T∨I∨P)	(T∨P∨L)	(P∨L)	(T∨I∨P∨L)	(T∨I∨P∨L)	(T∨I∨P)	(T∨I∨L)	φ

value is (T∨I∨L), which means that objects x1 and x3 are discernible with respect to attributes 'T', 'I' or 'L'. Note that diagonal is empty otherwise, it will repeat the entries.

3.3.7   Discernibility Function

A discernibility function f_A for an information system A is Boolean function such that

$$f_A = \cap\{ \cup c_{ij}|c_{ij} \neq \emptyset\} \tag{3.7}$$

Discernibility matrix combined with discernibility matrix help us in defining Reducts and rule extraction. We will see the example of rule extraction in the upcoming section. Consider the information system given in the table above, the discernibility function for this system can be defined as follows:

f_A(T,I,P,L) = (I∨L)(T∨I∨L) (T∨I∨L) (T∨L)(T∨I) (I∨P∨L) (T∨I∨P)
(T∨L) (T∨I) (T∨I) (T∨I∨L) (I∨P∨L) (T∨P∨L)
(T∨I∨L) (T∨I∨L) (T∨I∨L) (T∨I∨P) (P∨L)
(I) (L) (T∨P∨L) (T∨I∨P∨L)
(I∨L) (T∨I∨P∨L) (T∨I∨P∨L)
(T∨P∨L) (T∨I∨P)
(T∨I∨L)

After solving, the function simplifies to IL (i.e. $I \cap L$). There is a relation between discernibility function and discernibility matrix, each row in discernibility function corresponds to a column in discernibility matrix. The entries in discernibility function discern the objects from other, e.g. the second last row indicates that object x7 differs from x8 with respect to attributes T, I and L. A discernibility function based on column K from discernibility matrix is called K-discernibility function and thus identifies the minimum set of Reducts required to discern X_k from other objects.

3.3.8 Decision-Relative Discernibility Matrix

Decision-relative discernibility matrix is special form of discernibility matrix $M^d(A) = c_{ij}^d$ assuming $c_{ij}^d = \emptyset$ if $d(x_i) = d(x_j)$ and $c_{ij}^d = c_{ij}$ otherwise. As we construct the discernibility function from discernibility matrix, similarly decision-relative discernibility function can be constructed from decision-relative discernibility matrix. Simplification of this function results in a set of all the decision-relative Reducts of A.

We will now explain this with an example. We will consider the same information system given in the table above. However, we have rearranged the rows to order them according to decision class. The resulting discernibility matrix is shown in table below. It is a symmetrical matrix with empty diagonal. Similarly, all the entries where decision is equal are also empty.

Table 3.8 Decision-relative discernibility matrix for decision system shown in Table 3.7.

Table 3.7 Decision system given in Table 3.6 rearrange according to decision class

	T	I	P	L	d
x1	1	1	1	2	1
x4	1	2	1	0	1
x6	1	2	1	2	1
x7	1	2	0	1	1
x2	1	0	1	0	0
x3	2	0	1	1	0
x5	1	1	1	0	0
x8	2	0	0	2	0

Table 3.8 Decision-relative discernibility matrix

	X1	X4	X6	X7	X2	X3	X5	X8
x1	Φ							
X4	Φ	Φ						
X6	φ	Φ	Φ					
X7	φ	φ	Φ	Φ				
X2	I,L	T,I	T,I,L	I,P,L	Φ			
X3	T,I,L	T,I,L	T,I,L	T,I,P	Φ	Φ		
X5	T,L	I,	I,L	T,I,P,L	Φ	Φ	Φ	
x8	T,I,P	T,I,P,L	T,I,P	T,I,L	φ	Φ	φ	Φ

From the definition of decision-relative discernibility matrix, it follows that considering one column gives us discernibility function that can be used to discern that object from others, e.g. consider the column of X1, it gives the discernibility function that can be used to discern object X_1 from others. Figures 3.3, 3.4, 3.5 and 3.6 (taken from [3]) show four types of indiscernibility.

Figure 3.3 shows the indiscernibility relation not related to a particular concept and decision attribute. These Reducts are minimal subsets that discern all the cases from each other up to some extent.

Figure 3.4 shows the indiscernibility relation relative to a decision attribute but not relative to a particular case. These Reducts are minimal set of attributes that provide the same classification as obtained by entire set of conditional attributes.

Figure 3.5 shows the indiscernibility relation relative to a case or object X but not relative to decision attribute. Reducts of such type are minimum subset of

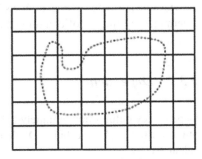

Fig. 3.3 Indiscernibility relation not related to a particular concept and decision attribute

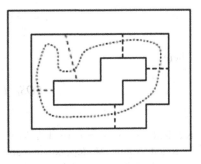

Fig. 3.4 Indiscernibility relation related to a decision attribute but not relative to a particular case

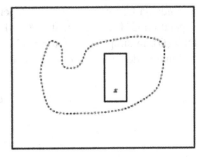

Fig. 3.5 Indiscernibility relation related to a case or object X but not relative to decision attribute

Fig. 3.6 Indiscernibility
relation related both to a case
as well as decision attribute

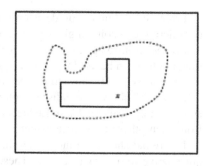

attributes that can discern object X from others up to the same extent as done by a
full set of conditional attributes.

Figure 3.6 shows the indiscernibility relation relative to both, i.e. a case as well
as decision attribute. These Reducts let us determine the outcome of a case as
determined by full set of conditional attributes.

3.3.9   Dependency

Dependency defines how uniquely the value of an attribute determines the value of
other attributes. An attribute 'D' depends on other attributes 'C' by degree 'K'
calculated by

$$k = \gamma(C, D) = \frac{|POS_C(D)|}{|U|} \tag{3.8}$$

where

$$POS_C(D) = \bigcup_{X \in U/D} \underline{C}(X) \tag{3.9}$$

is called positive region of 'U/D' with respect to 'C' as discussed in Sect. 3.1.5. 'K'
is called degree of dependency and specifies the ratio of the elements that can
positively be contained by partition induced by D, i.e. U/D. If K = 1, D fully
depends on C, for 0 < K < 1, D depends partially on C and for K = 0, D does not
depend on C. It is clear that if K = 1, i.e. D totally depends on C then IND(C) ⊆
IND (D), in simple words the U/C is finer than U/D.

Finally, dependency is calculated as follows:

$$k = \gamma(C, D) = \frac{|POS_C(D)|}{|U|} \tag{3.10}$$

Table 3.9 Sample decision system

U	State	Qualification	Job
x_1	S1	Doctorate	Yes
x_2	S1	Diploma	No
x_3	S2	Masters	No
x_4	S2	Masters	Yes
x_5	S3	Bachelors	No
x_6	S3	Bachelors	Yes
x_7	S3	Bachelors	No

Calculating dependency using positive region requires the following three steps.

1. First, construct the equivalence class structure using decision classes.
2. Construct equivalence class structure using the current attribute set.
3. Calculate positive region using:

Here, we provide details of each of these steps. Consider the decision system $DS = \{\{State, Qualification\} \cup \{Job\}\}$ given in Table 3.9:

We will calculate $k = \gamma(\{state, Qualification\}, Job)$ using positive region-based approach.

Step-1:

First step is calculating positive region-based dependency measure is to calculate equivalence classes using decision attribute ('Job' in our case):

Equivalence class structure specifies all the indiscernible objects, i.e. the objects which with respect to given attributes cannot be distinguished. In our case we will have two equivalence classes as follows:

$Q1 = \{x_1, x_4, x_6\}$
$Q2 = \{x_2, x_3, x_5, x_7\}$

Note that if consider the value of 'Job' as 'Yes' we cannot distinguish among x_1, x_4 and x_6.

Step-2:

After calculating the equivalence classes using decision attribute, the next step is to calculate equivalence class structure for decision attributes (in our case '{State, Qualification}'). Calculating equivalence classes using conditional attributes requires comparison of the value of each attribute for each record to find indiscernible objects. The equivalence classes in our case will be

$P1 = \{x1\}$
$P2 = \{x2\}$
$P3 = \{x3, x4\}$
$P4 = \{x5, x6, x7\}$

Step-3:

Positive region specifies which equivalence classes in Step-2 are contained by or subset of equivalence classes identified in Step-1. First, we will check which classes from P1...P4 are subset of Q1 and then we will calculate which classes from P1... P4 are subsets of Q2. This process will be used for all classes in Step-2 and we will identify all classes that are subset of equivalence classes in Step-1.

Here,

$$P_1 \subseteq Q_1$$

$$P_2 \subseteq Q_2$$

No other class from P_1, P_2, P_3 and P_4 is subset of either of Q_1 and Q_2. So the dependency will be

$$k = \gamma(C, D) = \frac{|POS_C(D)|}{|U|} = \frac{|P_1| + |P_2|}{|U|}$$

$$k = \gamma(\{State, Qualification\}, Job) = \frac{2}{7}$$

This process will take a considerable amount of time for datasets with large numbers of attributes and instances. Thus, this factor makes a positive region-based dependency measure a bad choice for use in feature selection algorithms against these datasets.

3.3.10 Reducts and Core

One way of dimensional reduction is keeping only those attributes that preserve the indiscernibility relation, i.e. classification accuracy. Using selected set of attributes provides the same set of equivalence classes that can be obtained by using the entire attribute set. The remaining attributes are redundant and can be reduced without affecting classification accuracy. There are normally many subsets of such attributes called Reducts. Mathematically, Reducts can be defined using the dependency as follows:

$$\gamma(C, D) = \gamma(C', D) \text{ for } C' \subseteq C \tag{3.11}$$

i.e. an attribute set C' $\subseteq$ C will be called Reduct with respect to D, if the dependency of D on C' will be same as that of its dependency on C.

Calculating the Reducts comprises of two steps. First, we calculate dependency of the decision attribute on entire dataset. Normally, this is '1', however, for inconsistent datasets, this may be any value between '0' and '1'. In the second step,

Table 3.10 Sample decision system

U	a	b	c	D
X_1	1	1	3	x
X_2	1	2	2	y
X_3	2	1	3	x
X_4	3	3	3	y
X_5	2	2	3	z
X_6	1	1	2	x
X_7	3	3	1	y

we try to find the minimum set of attributes on which decision attribute has the same dependency value as that of its value on entire set of attributes. In this step, we may use any Rough Set-based feature selection algorithm. It should be noted that there may be more than one Reduct sets in a single dataset.

We will now explain it with the help of an example. Consider Table 3.10 given below.

For our first step, we calculate dependency of decision attribute 'D' on conditional attributes $C = \{a,b,c\}$. Here we find that

$$\gamma(C,D) = 1$$

For the second step, we have to find the attribute subsets such that condition mentioned in Eq. (3.8) is satisfied. Here, we see that we may have two subsets that satisfy the condition.

$$\gamma(\{a, b\},D) = 1$$
$$\gamma(\{b, c\},D) = 1$$

Representing them with R_1 and R_2:

$$R_1 = \{a, b\}$$
$$R_2 = \{b, c\}$$

So either of R1 and R2 provides the same classification accuracy as provided by entire of the conditional attribute set thus can be used to represent entire dataset. It is important that Reduct set should be optimal, i.e. it should contain minimum number of attributes to better realize its significance, however, finding optimal Reduct is a difficult task as it requires exhaustive search with more number of resources. Normally exhaustive algorithms are used to find Reducts in smaller datasets, however, for datasets beyond smaller size, the other category of algorithms, i.e. random or heuristics based search are used, but the drawback of these algorithms is that they do not produce an optimal result. So getting the optimal Reducts is a trade-off between the resources and Reduct size.

Core is another important concept in Rough Set Theory. Normally, the Reduct set is not unique in a dataset, i.e. we may have more than one Reduct sets. Although Reduct may contain the same amount of information otherwise represented by entire attribute subset, but even in Reduct there are attributes that are more important than others, i.e. these attributes cannot be removed without affecting the classification accuracy of the Reducts. Mathematically, it can be written as $Core = \bigcap_{i=1}^{n} R_i$ where R_i is ith Reduct Set. So, core is the attribute or set of attributes common to all Reduct sets. In our example explained above it is clear that the attribute {b} is common in all Reduct sets, so, {b} is core attribute here. Manually it can be seen that removing attribute {b} from either of the Reducts effects dependency of decision class on rest of the attributes in that Reduct and thus effecting the classification accuracy of the Reduct.

3.4   Discretization Process

Discretization is the process of converting or partitioning continuous attributes, features or variables to discretized or nominal attributes/features/variables/intervals. Discretization process determines, how coarsely we see the world, e.g. student marks can be any real number between zero (0) to hundred (100) in a subject but we normally convert it to three to four grades. Another example may be blood pressure. Although the value is already measured in discrete (natural number), however, it may be determined in three intervals. So, it is apparent that discretization of features is a complex step, however, it is essential especially from feature selection point of view as the data in real life may be continuous. So before performing feature selection, we have to apply discretization step to convert the attribute values to discrete values.

Various techniques have been proposed in literature for discretization [46, 68]. Here, we will present a simple RST based discretization method taken from [3]. In discretization of basic decision system $\Delta = (U, A \cup \{d\})$, where $V_a = [v_a, \omega_a)$ is an interval of real numbers, we try to find partition P_a of V_a for $a \in A$. Any partition of V_a is defined by a sequence of cuts $V_1 < V_2 < \ldots < V_n$ from V_a. So in discretization, we identify some family of partitions by finding the cuts that satisfy natural condition.

Following are the sequence of steps in an RST discretization process taken from [3].

Step-1: For each attribute, create a set c(U) where $a \in A$.
Step-2: Determine $V_a = [v_a, \omega_a)$ for each attribute $a \in A$.
Step-3: Define intervals on value sets defined in Step-2
Step-4: Create a set of Cuts P for each interval
Step-5: Discretize the attribute a to new attribute a^P using the cuts.

The process steps are defined at abstract level, now we explain all of these steps with the help of an example.

Example:

Consider Table 3.10 given below. It comprises of two conditional attributes and one decision attribute. Conditional attributes 'a' and 'b' have real values.

The first step is to determine $V_a = [v_a, \omega_a)$ for each attribute $a \in A$. The values of attributes 'a' and 'b' for universe U are given by the sets:

a(U) = {0.8,1,1.3,1.4,1.6} and b(U) = {0.5,1,2,3}

Here we have two attributes, so, $V_a = [0.8, 2)$ and $V_b = [0.5, 4)$.

The Next step is to define intervals based on the value set of conditional attributes, in our case:

For attribute 'a': [0:8:1); [1; 1:3); [1:3; 1:4); [1:4; 1:6);

For attribute 'b': [0:5; 1); [1; 2); [2; 3);

The next step is to form cuts. A cut is a pair (a,c) where $a \in A$. A cut may simply be a middle point of the intervals defined above. So,

In case of attribute 'a': P_a = (a; 0:9); (a; 1:15); (a; 1:35); (a; 1:5);

In case of attribute 'b': P_b = (b; 0:75); (b; 1:5); (b; 2:5);

Figure 3.7 given below shows the graphical representation of intervals and cuts.

Figure 3.8 (below) shows the relation between data and cuts. These sets of cuts define new attribute a_p for any value of a. This mechanism works as follows:

The value of attribute may fall anywhere in between values of cuts. We suppose the cut set P_a = (a; 0.9); (a; 1.15); (a; 1.35); (a; 1.5); Now any value from V_a that falls below 0.9 will be given value of '1', any value that will fall in range between 0.9 and 1.5 will be given value of '1', any value in range 1.15 to 1.35 will be given value of '2' and so on. Similarly for P_b = (b; 0.75); (b; 1.5); (b; 2.5); any value below 0.75 will be assigned a value 0 and any value from 0.75 to 1.5 will be assigned value of '1' and so on.

So using the Table 3.10 will be discretized as follows (given in Table 3.11).

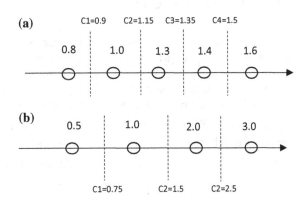

Fig. 3.7 Graphical representation of intervals and cuts

Fig. 3.8 Graphical
representation of relation
between data and cuts

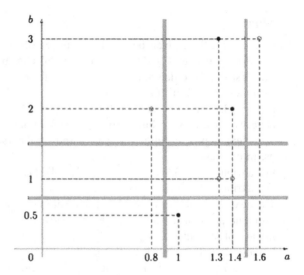

Table 3.11 A decision
system comprising of real
numbers

A	a	b	d
u1	0.8	2	1
u2	1	0.5	0
u3	1.3	3	0
u4	1.4	1	1
u5	1.4	2	0
u6	1.6	3	1
u7	1.3	1	1

3.5   Miscellaneous Concepts

Now we will discuss some miscellaneous concepts taken from [3]. For a decision
system $A = (U, C \cup \{d\})$, the cardinality of the image d(U) is called rank of
decision attribute 'd' and is denoted by r(d). For example, for decision system in
Table 3.2 rank of decision class is '2'.

A decision system can also be partitioned on the basis of decision attribute.
$CLASS_A(d) = \{X_A^1, X_A^2, \ldots, X_A^{r(d)}\}$ is called classification of objects in A determined by d, where X_A^i is called ith decision class.

If $\{X_A^1, X_A^2, \ldots, X_A^{r(d)}\}$ are decision classes of decision system A, then
$\left\{\underline{C}_{x1} \cup \underline{C}_{x2} \ldots \underline{C}_{r(d)}\right\}$ is called positive region.

Fig. 3.9 Graphical representation of few RST application

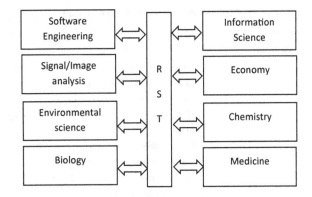

For decision system given in Table 3.2, there are two decision classes, i.e. {1,0}. The partitioning of the universe for this table using decision attribute is $U = \{X^1 \cup X^2\}$ where $X^1 = \{u1, u4, u6, u7\}$ and $X^0 = \{u2, u3, u5\}$.

For a decision system $A = (U, C \cup \{d\})$, the decision system is consistent if and only $POS_A(d) = U$.

3.6 Applications of RST

Right from its inception, it has been used in various domains for data analysis including economy and finance [4], medical diagnosis [5], medical imaging [6], banking [7], data mining [8], etc. Figure 3.9 shows the graphical representation of different applications of RST.

Here, we present some of the representative applications of RST in different domains. Table 3.12 shows sample applications of RST in different domains (Table 3.13).

Table 3.12 Discretized decision system

A	a	b	d
u1	0	2	1
u2	1	0	0
u3	1	2	0
u4	1	1	1
u5	1	2	0
u6	2	2	1
u7	1	1	1

Table 3.13 Applications of RST in different domains [3]

Domain	Application	References
Medicine	Treatment of duodenal ulcer by HSV	[9–13]
	Analysis of data from peritoneal lavage in acute pancreatitis	[13, 14]
	Knowledge acquisition in nursing	[15]
	Diagnosis of pneumonia patients	[16]
	Medical databases (e.g. headache, meningitis, CVD) analysis	[17]
	Image analysis for medical applications	[18]
	Surgical wound infection	[19]
	Classification of histological pictures	[20]
	Preterm birth prediction	[19]
	Verification of indications for treatment of urinary stones by extra corporeal shock wave lithotripsy (ESWL)	[21]
	Analysis of factors affecting the differential diagnosis between viral and bacterial meningitis	[22, 23]
	Developing an emergency room for diagnostic check list A case study of appendicitis	[24]
	Analysis of medical experience with urolithiasis patients treated by extra-corporeal shock wave lithotripsy	[25]
	Diagnosing in progressive encephalopathy	[26–28]
	Rough set-based filtration of sound applicable to hearing prostheses	[29]
	Discovery of attribute dependencies in experience with multiple injured patients	[30]
	Modelling cardiac patient set residuals	[31]
	Multistage analysis of therapeutic experience with acute pancreatitis	[32]
	Breast cancer detection using electro-potentials	[33]
	Analysis of medical data of patients with suspected acute appendicitis	[34]
	Attribute reduction in a database for hepatic diseases	[35]
	EEG signal analysis	[36]
Economics finance and business	Evaluation of bankruptcy risk	[37–39]
	Company evaluation	[40]
	Customer behaviour patterns	[41]
	Response modelling in database marketing	[42]
	Analysis of factors affecting stock price fluctuation	[43]
	Discovery of strong predictive rules for stock market	[44]
	Purchase prediction in database marketing	[45]
	Modelling customer retention	[46]
	Temporal patterns	[47]
	Analysis of business databases	[48]
		[49]

(continued)

Table 3.13 (continued)

Domain	Application	References
	Rupture prediction in a highly automated production system	
Environmental cases	Analysis of a large multispecies toxicity database	[50]
	Drawing premonitory factors for earthquakes by emphasizing gas geochemistry	[51]
	Control conditions on a polder	[52]
	Global warming: influence of different variables on the earth global temperature	[53]
	Global temperature stability	[54]
	Programming water supply systems	[55, 56]
	Predicting water demands in Regina	[57]
	Prediction of slope-failure danger level from cases	[58]
Signal and image analysis	Noise and distortion reduction in digital audio signal	[59, 60]
	Filtration and coding of audio	[61]
	Recognition of musical sounds	[62]
	Detection and interpolation of impulsive distortions in old audio recordings	[63]
	Subjective assessment of sound quality	[64]
	Classification of musical timbres and phrases	[65]
	Image analysis	[18]
	Converting a continuous tone image into a half-tone image using error diffusion and rough set methods	[66]
	Voice recognition	[29]
	Handwritten digit recognition	[67]
Software Engineering	Qualitative analysis of software engineering data	[68]
	Assessing software quality	[69]
	Software deployability	[70]
	Knowledge discovery form software engineering data	[71]
Information sciences	Information retrieval	[72]
	Analysis and synthesis of concurrent systems	[73]
	Integration RDMS and data mining tools using rough sets	[74]
	Rough set model of relational databases	[75]
	Cooperative knowledge base systems	[76]
Molecular biology	Discovery of functional components of proteins from amino acid sequences	[77]
Chemistry: pharmacy	Analysis of relationship between structure and activity of substances	[78]

3.7   Summary

This chapter has in-depth discussed the preliminary concepts of Rough Set Theory. To start RST, it was necessary to have some basic concepts of classical set theory. So, for this purpose, some basic concepts of classical Set Theory were also discussed. For RST each and every concept was complemented with an example to help the reader grasp the concept. We have also provided details of some miscellaneous topics including discretization process and some other definitions.

A thorough literature review regarding the use of RST in various domains along with its exact application has also been made part of the chapter. Special efforts were made to keep the presentation method simple and basic, so that it could help the readers understand RST that has always been a tough task to learn due to its mathematical nature. In the upcoming chapter, we will discuss some advanced topics in RST.

References

1. http://www.rapidtables.com/math/symbols/Basic_Math_Symbols.htm. Access 30 Mar 2017
2. Pawlak Z (1991) Rough sets: theoretical aspects of reasoning about data. Kluwer Academic, Dordrecht
3. Pal SK, Skowron A (1999) Rough-fuzzy hybridization: a new trend in decision making. Springer, New York Inc
4. Krysiński J (1990) Rough sets approach to the analysis of the structure-activity relationship of quaternary imidazolium compounds. Arzneimittelforschung 40(7):795–799
5. Podsiadło M, Rybiński H (2014) Rough sets in economy and finance. Transactions on Rough Sets XVII. Springer, Berlin Heidelberg, pp 109–173
6. Prasad V, Srinivasa Rao T, Surendra Prasad Babu M (2016) Thyroid disease diagnosis via hybrid architecture composing rough data sets theory and machine learning algorithms. Soft Comput 20(3):1179–1189
7. Xie C-H, Liu Y-J, Chang J-Y (2015) Medical image segmentation using rough set and local polynomial regression. Multimed Tools Appl 74(6):1885–1914
8. Montazer GA, ArabYarmohammadi S (2015) Detection of phishing attacks in Iranian e-banking using a fuzzy–rough hybrid system. Appl Soft Comput 35:482–492
9. Pawlak Z, Słowiński K, Słowiński R (1986) Rough classification of patients after highly selective vagotomy for duodenal ulcer. Int J Man Mach Stud 24(5):413–433
10. Fibak J et al (1986) Rough sets based decision algorithm for treatment of duodenal ulcer by HSV. Biol Sci 34:227–249
11. Fibak J, Slowinski K, Slowinski R (1986) The application of rough set theory to the verification of indications for treatment of duodenal ulcer by HSV. In: Proceedings 6th internat, workshop on expert systems and their applications, Avignon, France, Vol 1, pp 587–599
12. Slowinski R, Slowi_nski K (1989) An expert system for treatment of duodenal ulcer by highly selective vagotomy (in Polish). Pamietnik 54. Jubil. ZjazduTowarzystwaChirurgow Polskich, Krakow I, 223–228
13. Slowinski K (1992) Rough classification of HSV patients. In: Intelligent decision support-handbook of applications and advances of the rough sets theory, pp 77–94

14. Słowiński K (1994) Rough sets approach to analysis of data of diagnostic peritoneal lavage applied for multiple injuries patients. In: Rough sets, fuzzy sets and knowledge discovery. Springer, London, pp 420–425

15. Słowiński K, Slnowiński R, Stefanowski J (1988) Rough sets approach to analysis of data from peritoneal lavage in acute pancreatitis. Med Inform 13(3):143–159

16. Grzymala-Busse JW (1998) Applications of the rule induction system LERS. Rough Sets in Knowledge Discovery 1, pp 366–375

17. Paterson GI (1994) Rough classification of pneumonia patients using a clinical database. Rough Sets, Fuzzy Sets and Knowledge Discovery. Springer, London, pp 412–419

18. Tsumoto S, Tanaka H (1995) PRIMEROSE: probabilistic rule induction method based on rough sets and resampling methods. Comput Intell 11(2):389–405

19. Jelonek J et al (1994) Neural networks and rough sets—comparison and combination for classification of histological pictures. Rough Sets, Fuzzy Sets and Knowledge Discovery. Springer, London, pp 426–433

20. Kandulski M, Marciniec J, Tukałło K (1992) Surgical wound infection—conducive factors and their mutual dependencies. Intelligent decision support. Springer, Netherlands, pp 95–110

21. Grzymala-Busse JW, Linda KG (1996) A comparison of less specific versus more specific rules for preterm birth prediction. In: Proceedings of the first online workshop on soft computing WSC1 on the internet, Japan

22. Slowinski K et al (1995) Rough set approach to the verification of indications for treatment of urinary stones by extracorporeal shock wave lithotripsy (ESWL). Soft Computing, Society for Computer Simulation, San Diego, California, pp 142–145

23. Tsumoto S, Ziarko W (1996) The application of rough sets-based data mining technique to differential diagnosis of meningoenchepahlitis. International Symposium on Methodologies for Intelligent Systems. Springer, Berlin, Heidelberg

24. Ziarko W (1998) Rough sets as a methodology for data mining. Rough Sets Knowl Discov 1:554–576

25. Rubin S, Michalowski W, Slowinski R (1996) Developing an emergency room diagnostic check list using rough sets-a case study of appendicitis. Simul Med Sci, 19–24

26. Slowinski K, Stefanowski J (1996) On limitations of using rough set approach to analyse non-trivial medical information systems

27. Paszek P, Wakulicz Deja A (1996) Optimalization diagnose in progressive encephalopathy applying the rough set theory. Zimmermann 557.1:192–196

28. Wakulicz-Deja A, Boryczka M, Paszek P (1998) Discretization of continuous attributes on decision system in mitochondrial encephalomyopathies. In: International conference on rough sets and current trends in computing. Springer, Berlin, Heidelberg

29. Wakulicz-Deja A, Paszek P (1997) Diagnose progressive encephalopathy applying the rough set theory. Int J Med Informatics 46(2):119–127

30. Czyzewski A (1998) Speaker-independent recognition of isolated words using rough sets. Inf Sci 104(1-2):3–14

31. Stefanowski J, Slowiński K (1997) Rough set theory and rule induction techniques for discovery of attribute dependencies in medical information systems. European Symposium on Principles of Data Mining and Knowledge Discovery. Springer, Berlin, Heidelberg

32. Ohrn A et al (1997) Modelling cardiac patient set residuals using rough sets. In: Proceedings of the AMIA annual fall symposium. American medical informatics association

33. Słowiński K, Stefanowski J (1998) Multistage rough set analysis of therapeutic experience with acute pancreatitis. Rough Sets in Knowledge Discovery 2. Physica-Verlag HD, pp 272–294

34. Swiniarski RW (1998) Rough sets and bayesian methods applied to cancer detection. In: International conference on rough sets and current trends in computing. Springer, Berlin, Heidelberg

35. Carlin US, Komorowski J, Øhrn A (1998) Rough set analysis of patients with suspected acute appendicitis. Traitement d'information et gestion d'incertitudes dans les systèmes à base de connaissances. Conférence internationale

36. Tanaka H, Maeda Y (1998) Reduction methods for medical data. Rough Sets in Knowledge Discovery 2. Physica-Verlag HD, pp 295–306
37. Wojdyłło P (1998) Wavelets, rough sets and artificial neural networks in EEG analysis. In: International conference on rough sets and current trends in computing. Springer, Berlin, Heidelberg
38. Slowinski R, Zopounidis C (1995) Application of the rough set approach to evaluation of bankruptcy risk. Intell Syst Account Finance Manag 4(1):27–41
39. Slowinski R, Zopounidis C (1994) Rough-set sorting of firms according to bankruptcy risk. Applying Multiple Criteria Aid for Decision to Environmental Management. Springer, Netherlands, pp 339–357
40. Greco S, Matarazzo B, Slowinski R (1998) A new rough set approach to evaluation of bankruptcy risk. Operational tools in the management of financial risks. Springer, US, pp 121–136
41. Mrózek A, Skabek K (1998) Rough sets in economic applications. Rough Sets in Knowledge Discovery 2. Physica-Verlag HD, pp 238–271
42. Piasta Z, Lenarcik A (1998) Learning rough classifiers from large databases with missing values. Rough Sets Knowl Discov 1:483–499
43. Van den Poel D (1998) Rough sets for database marketing. Rough Sets in Knowledge Discovery 2. Physica-Verlag HD, pp 324–335
44. Golan RH, Ziarko W (1995) A methodology for stock market analysis utilizing rough set theory. In: Computational intelligence for financial engineering, Proceedings of the IEEE/ IAFE. IEEE
45. Ziarko W, Golan R, Edwards D (1993) An application of datalogic/R knowledge discovery tool to identify strong predictive rules in stock market data. In: Proceedings of AAAI workshop on knowledge discovery in databases, Washington, DC
46. Van den Poel D, Piasta Z (1998) Purchase prediction in database marketing with the ProbRough system. In: International conference on rough sets and current trends in computing. Springer, Berlin, Heidelberg
47. Kowalczyk AE, Eiben TJ, Euverman W, Slisser F (1999) Modelling customer retention with statistical techniques, rough data models, and genetic programming. Rough Fuzzy Hybridization: A New Trend in Decision-making
48. Kowalczyk W (1996) Analyzing temporal patterns with rough sets. Zimmermann 557:139
49. Kowalczyk W, Piasta F (1998) Rough-set inspired approach to knowledge discovery in business databases. In: Pacific-Asia conference on knowledge discovery and data mining. Springer, Berlin, Heidelberg
50. Swiniarski R et al (1997) Feature selection using rough sets and hidden layer expansion for rupture prediction in a highly automated production process. Syst Sci-Wroclaw- 23:53–60
51. Keiser K, Szladow A, Ziarko W (1992) Rough sets theory applied to a large multispecies toxicity database. In: Proceedings of the Fifth international workshop on QSAR in environmental toxicology, Duluth, Minnesota
52. Teghem J, Charlet J-M (1992) Use of "Rough Sets" method to draw premonitory factors for earthquakes by Emphasing gas geochemistry: the case of a low seismic activity context, in Belgium. Intelligent Decision Support. Springer, Netherlands, pp 165–179
53. Reinhard A et al (1992) An application of rough set theory in the control conditions on a polder. S lowi nski 428:331
54. la Busse, Grzyma JW, Gunn JD (1995) Global temperature analysis based on the rule induction system LERS. In: Proceedings of the fourth international workshop on intelligent information systems, August ow, Poland, June. Vol 5. No 9
55. Gunn JD, Grzymala-Busse JW (1994) Global temperature stability by rule induction: an interdisciplinary bridge. Human Ecology 22(1):59–81
56. Greco S, Matarazzo B, Słowiński R (1998) "Rough approximation of a preference relation in a pairwise comparison table. Rough Sets in Knowledge Discovery 2. Physica-Verlag HD, pp 13–36

57. Roy B, Slowinski R, Treichel W (1992) "Multicriteria programming of water supply systems for rural AREAS1, 13–31
58. An A et al (1995) Discovering rules from data for water demand prediction. In: Proceedings of the workshop on machine learning in engineering IJCAI. Vol 95
59. Furuta H, Hirokane M, Mikumo Y (1998) "Extraction method based on rough set theory of rule-type knowledge from diagnostic cases of slope-failure danger levels. Rough Sets in Knowledge Discovery 2. Physica-Verlag HD, pp 178–192
60. Czyzewski A (1996) Mining knowledge in noisy audio data. KDD
61. Czyżewski A (1998) Soft processing of audio signals. Rough Sets in Knowledge Discovery 2. Physica-Verlag HD, pp 147–165
62. Czyzewski A, Krolikowski R (1997) New methods of intelligent filtration and coding of audio. Audio Engineering Society Convention 102. Audio Engineering Society
63. Kostek B (1998) Soft computing-based recognition of musical sounds. Rough Sets in Knowledge Discovery 2. Physica-Verlag HD, pp 193–213
64. Czyzewski A (1995) Some methods for detection and interpolation of impulsive distortions in old audio recordings. In: IEEE ASSP workshop on applications of signal processing to audio and acoustics. IEEE
65. Kostek B (1998) Soft set approach to the subjective assessment of sound quality. In: Fuzzy systems proceedings, 1998. IEEE world congress on computational intelligence, The 1998 IEEE International Conference on. Vol 1. IEEE
66. Kostek B, Szczerba M (1996) Parametric representation of musical phrases. Audio Engineering Society Convention 101. Audio Engineering Society
67. Zeng H, Swiniarski R (1998) A new halftoning method based on error diffusion with rough set filtering. Rough Sets in Knowledge Discovery 2. Physica-Verlag HD, pp 336–342
68. Bazan JG et al (1998) Synthesis of decision rules for object classification. Incomplete Information: Rough Set Analysis. Physica-Verlag HD, pp 23–57
69. Ruhe G (1996) Qualitative analysis of software engineering data using rough sets. Tsumoto, Kobayashi, Yokomori, Tanaka, and Nakamura 484:292
70. Peters JF, Ramanna S (1999) A rough sets approach to assessing software quality: Concepts and rough Petri net models. Rough-Fuzzy Hybridization: New Trends in Decision Making. Springer, Berlin, pp 349–380
71. Peters JF, Ramanna S (1998) Software deployability decision system framework: a rough set approach. Traitement d'information et gestion d'incertitudes dans les systèmes à base de connaissances. Conférence internationale
72. Ruhe G (1997) Knowledge discovery from software engineering data: Rough set analysis and its interaction with goal-oriented measurement. European Symposium on Principles of Data Mining and Knowledge Discovery. Springer, Berlin, Heidelberg
73. Srinivasan P (1989) Intelligent information retrieval using rough set approximations. Inf Process Manage 25(4):347–361
74. Skowron A, Suraj Z (1993) Rough sets and concurrency. Bull Polish Acad Sci. Technical sciences 41.3:237–254
75. Nguyen SH et al (1996) Knowledge discovery by rough set methods. In: Proceedings of the international conference on information systems analysis and synthesis ISAS. Vol 96
76. Beaubouef T, Petry FE (1994) A rough set model for relational databases. Rough Sets, Fuzzy Sets and Knowledge Discovery. Springer, London, pp 100–107
77. Ras ZW (1996) Cooperative knowledge-based systems. Intell Autom Soft Comput 2(2):193–201
78. Tsumoto S, Tanaka H (1995) Automated discovery of functional components of proteins from amino-acid sequences based on rough sets and change of representation. KDD
79. Maciá-Pérez F et al (2015) Algorithm for the detection of outliers based on the theory of rough sets. Decis Support Syst 75:63–75

Chapter 4
Advanced Concepts in Rough Set Theory

In the last chapter, we discussed some basic concepts of Rough Set Theory. In this chapter, we will present some advanced concepts including some improved definitions, their examples and hybridization of RST with fuzzy set theory.

4.1 Fuzzy Set Theory

Later in this chapter, we will discuss a few concepts related to hybridization of RST with fuzzy set theory, but we need to first explain some basic concepts of Fuzzy Set Theory as discussed below.

4.1.1 Fuzzy Set

If X is a universe of disclosure and x is a particular element of X, then a fuzzy set F defined on X will be a collection of ordered pairs:

$$A = \{(x, \mu_A(x)), x \in X\} \tag{4.1}$$

Where each pair $(x, \mu_F(x))$ is called a singleton. In case of crisp set theory, we don't use $\mu_F(x)$ because for all the elements which are present their degree would be one and for non-present elements degree would be zero. Consider the following set U:

$$U = \{0, 1, 2, 3, 4, 5, 6, 7, 8, 9\}$$

$$A = \{1, 3, 5, 7, 9\}$$

© Springer Nature Singapore Pte Ltd. 2019
M. S. Raza and U. Qamar, *Understanding and Using Rough Set
Based Feature Selection: Concepts, Techniques and Applications*,
https://doi.org/10.1007/978-981-32-9166-9_4

$$A = \{(0,0),(1,1),(2,0)(3,1),(4,0),(5,1),(6,0),(7,1),(8,0)(9,1)\}$$

Here note that we have shown ten ordered pairs, one for each element. Each ordered pair shows the element and its degree of presence in the set. For example, the first pair (0,1) shows that degree of presence of element '0' is '0', i.e. it is present in the set A, on the other hand, the pair (1,1) shows that element '1' is present in 'A'. It should be noted that the degree of an element has the range from zero (0) to one (1). For example, consider the reference set A of intelligent students, where 'intelligent' is a fuzzy term:

$$A = S1, S2, S3, S4, S5$$

Now it should be noted that each person has his/her own intelligence level. It is not that a student will have zero intelligence level while others will have a level of 100%. So, a fuzzy set may take the form:

$$A = \{(S1,0.1),(S2,0.4),(S3,0.5)(S4,0.5),(S5,1)\}$$

The above set shows the degree or strength, a student belongs to Set A. Student S2 is more intelligent than S1 similarly S5 is more smart than S4. Table 4.1 shows the difference between fuzzy set and concrete set. Figures 4.1 and 4.2 show the graphical representation of both.

Crisp set shows that only Saturday and Sunday will be part of the Set of Weekend days, while in the fuzzy set 'Friday' may also have partial membership, this is because we normally begin to feel weekend on Friday evening.

4.1.2 Fuzzy Sets and Partial Truth

As discussed earlier, this universe is not crisp, we come across many situations in real-life scenarios where we find the partial answers, e.g. if you ask about how much work you have done you will probably get an answer like almost, an only little bit left. Now the question is that what is meant by almost either it means all of the work or 60, 70% or how much. So the fuzzy logic gives us the ability to answer a question with a partial degree of certainty in formation. Fuzzy logic thus generalizes the Yes/No or 1/0 logic, i.e. in fuzzy logic, the answer maybe 0.72 or 0.5

Table 4.1 Difference between fuzzy set and concrete set

Fuzzy	Classic
Do not restrict some member to be fully part of a set	Restricts an object to either be member of a set or not
Allow partial membership	Don't allow partial membership
If we consider fuzzy set of weekends, Friday may be partial member of set	Only Saturday and Sunday will be members

Fig. 4.1 Fuzzy versus Crisp set

Fig. 4.2 Fuzzy versus Crisp set

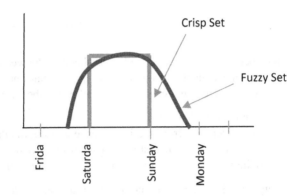

instead of exact 1 or 0. So in fuzzy logic truth has its degree, a value indicating the amount of truth in a statement, e.g. consider the following questions and their answers:

Q   Are you in Seventh class?
Answer   No (Crisp)
Q   Are you going to New York?
Answer   Yes (Crisp)
Question How much you like to go camping on weekends
Answer   Not too much (Fuzzy)
Question How much you are excited on weekends
Answer   Very much (Fuzzy)

If you note the last two questions, the answer is somewhere between 'Yes' and 'No' in contrast with the first two questions.

4.1.3 Membership Function

The degree of truth that we have talked about is determined by a membership function. A membership function is a graph that determines the elements of the

Fig. 4.3 Fuzzy versus Crisp
set

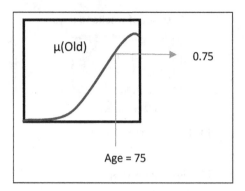

input space to the degree of truth (membership) between 0 and 1. It can range from
a simple linear function to a complex polynomial one.

Following are some characteristics of a membership function:

- It is represented by symbol μ.
- It takes elements from the input space and returns their degree of membership.
- The degree of membership ranges from zero to one, where zero represents
  absolute falseness and one represents absolute truth.

Figure 4.3 represents a membership functions μ(Old) that determines the degree
of membership (truth) of a person's age to the set of old people:

So, if a person's age is 75 years, the membership function μ(Old) returns
$\mu = 0.75$, which means that the person is quite old. Note that 0.75 just provides
information about the tendency of this person to belong to the set of old persons.

4.1.4   Fuzzy Operators

Before we move on to fuzzy operators, let's take a look at conventional logical
AND, OR and NOT operators (Fig. 4.4).

The table provides the Boolean input and corresponding crisp values after
applying the AND, OR and NOT operators. Fuzzy operators, on the other hand, can
act on real numbers as well. In fuzzy set theory, we use Min (minimum), Max
(Maximum) and 1-A for AND, OR and NOT operators, respectively. The terms
Intersection, Union and Complements are also used.

4.1.4.1   Union

Let A and B be two fuzzy sets, If μ(A) and μ(B) are two membership functions on
universe X, then union of μ(A) and μ(B) can be defined by fuzzy union operator as
follows:

AND Operator   **OR Operator**   **NOT Operator**

A	B	A AND B
0	0	0
0	1	0
1	0	0
1	1	1

A	B	A OR B
0	0	0
0	1	1
1	0	1
1	1	1

A	NOT A
0	1
1	0

Fig. 4.4 Conventional logical operators

$$\mu_{A \cup B}(X) = MAX(\mu(A), \mu(B)) \qquad (4.2)$$

Example:

Table 4.2 explains the union operator for two fuzzy sets A and B.

4.1.4.2 Intersection

Let A and B be two fuzzy sets, If $\mu(A)$ and $\mu(B)$ are two membership functions on universe X, then union of $\mu(A)$ and $\mu(B)$ can be defined by fuzzy intersection operator as follows:

$$\mu_{A \cap B}(X) = MIN(\mu(A), \mu(B)) \qquad (4.3)$$

Table 4.3 explains the intersection operator for two fuzzy sets A and B.

Table 4.2 Union operator

$\mu(A)$	$\mu(B)$	$\mu_{A \cup B}(X) = MAX(\mu(A), \mu(B))$
0.5	0.2	0.5
0.1	0.05	0.1
0.0	0.0	0.0
0.3	0.35	0.35

Table 4.3 Intersection operator

$\mu(A)$	$\mu(B)$	$\mu_{A \cup B}(X) = MIN(\mu(A), \mu(B))$
0.5	0.2	0.2
0.1	0.05	0.05
0.0	0.0	0.0
0.3	0.35	0.3

4.1.4.3 Complement

Let A be a fuzzy set and μ(A) is a membership function on universe X, then complement operator is defined as

$$\mu_{A^c}(X) = 1 - \mu(A) \tag{4.4}$$

The following Table 4.4 explains the complement operator for the fuzzy set A. Figure 4.5 shows all the three operators.

4.1.5 Fuzzy Set Representation

A fuzzy set can be represented in two ways. First, by using a triangular graph, the same as given in the figure above where the peak represents the means value. This is based on the assumption that most of the population lies at average or mean value, whereas exceptional cases are represented by the far edges on the slope of the triangle. The other and most convenient way to represent fuzzy sets is to represent a set in the form of value and membership pair. For example, to represent a fuzzy set 'Tall' about the height of people we can use the following notation.

$$Tall = 0/4, 0/4.5, 0/5, 0.25/5.5, 0.5/6, 0.75/6.5, 1/7$$

Here numerator represents the membership value and the denominator represents the actual height. So a person with the height up to 4.5 inches will not be considered tall whereas a person with height 7 inch will definitely be considered as tall.

Table 4.4 Complement operator

μ(A)	$\mu_{A^c}(X) = 1 - \mu(A)$
0.5	0.5
0.1	0.9
0.0	1.0
0.3	0.7

Union Operator

Intersection Operator

Complement Operator

Fig. 4.5 Fuzzy operators

4.1.6 Fuzzy Rules

Normally rules are implemented in the form of IF–Else Statements, e.g. from Table 4.5, we can have two rules as follows:

If X = A THEN Y = 0

This was a simple rule, we can have more complicated rules as well, e.g. from the above table, we can also derive that

If (X = B) OR (X = C) THEN Y = 1

Here, X and Y are variables and {B, C, 0, 1} are fuzzy distributions or sets. A more real-life rule may be

If slop is steep THEN energy consumption is high

Here the variable slop, in the fuzzy system may have crisp values, e.g. some number from 1 to 5. A rule has two parts called antecedents and consequent.

Antecedents are the conditions that need to be fulfilled to get the result, e.g. if weather is good then we can go out for camping. As discussed above, antecedents may have multiple parts concatenated using logical operators, in which case all parts are resolved to some single number.

Consequent is the resultant part, i.e. the portion after the word 'THEN'. Consequent may also have multiple parts, e.g. in the following rule:

If slop is steep THEN energy consumption is high and speed is slow.

We have two consequent parts. Antecedents affect consequents. We use the implication function to modify the output fuzzy set to the degree specified by antecedents. For example, for the rule:

If Slop is Steep OR Road is Rough THEN Energy Consumption is High.

The fuzzy system works as given in Fig. 4.6. The process comprises three steps:

Step-1: Fuzzify the inputs

All the antecedents are solved to a single value, the degree of membership between 0 and 1. For example, in the above-mentioned rule, the slope was given a value

Table 4.5 A raw dataset with two variables	X	Y
	A	0
	B	1
	A	0
	C	1

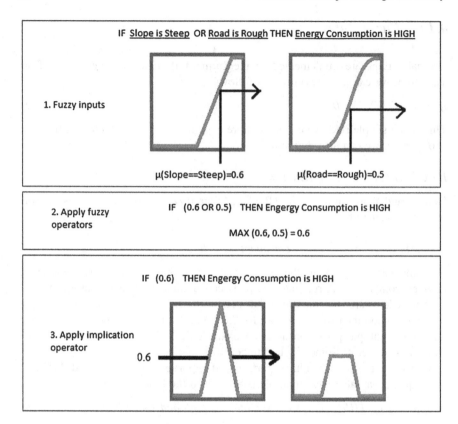

Fig. 4.6 How fuzzy system works

60% which fuzzifies to 0.6 by fuzzy membership function in the fuzzy set of high slope. Similarly, the roughness of the road was given a value 50% which resolved to 0.5 in the fuzzy set of Roughness.

Step-2: Apply Fuzzy operators

If rule comprises of multiple antecedents, then fuzzy operators are applied to solve the antecedent values. In our example, there are two antecedents, i.e. 'Slope' and 'Road'. There is an OR operator between them so max(0.6, 0.5) results in 0.6 called degree of support for rule.

Step-3: Applying the implication method

Apply the implication method and use the degree of support to shape the output fuzzy set. The consequent of a fuzzy rule assigns an entire fuzzy set to the output.

This fuzzy set is represented by a membership function that is chosen to indicate the qualities of the consequent. Then the output fuzzy set is truncated according to the implication method (Fig. 4.6).

4.2   Fuzzy Rough Set Hybridization

Both fuzzy and rough set theory are important components of computations. Rough set theory models uncertainty whereas fuzzy sets model vagueness. Researchers have already explored a number of ways in which both theories can interact with each other [1]. Here, we will first see some of the applications of fuzzy rough hybridization in supervised learning, information retrieval and feature selection.

4.2.1   Supervised Learning and Information Retrieval

In [2], the authors have enhanced the classification accuracy of K-nearest neighbours algorithm using fuzzy-rough uncertainty, however, still maintaining the simplicity and nonparametric characteristics. In the proposed solution, we do not need to know the value of 'K' as in the case of conventional one. Furthermore, the class confidence values calculated using fuzzy-rough ownership values do not sum up to one which helps algorithm distinguish between equal evidence and ignorance, and thus make the semantics of the class confidence values richer.

Natural languages contain a lot of uncertainty and vagueness, so fuzzy-rough hybridization can better help us solve natural language related problems. In [3], the authors have used fuzzy-rough framework for query refinement. They define a thesaurus using upper approximation where the query is approximated from both upper and lower side. The upper approximation related in query explosion whereas the lower approximation was found to be too strict resulting in the empty query. So the authors propose the use of lower approximation of the upper approximation (different from upper approximation) in the case when thesaurus is not transitive.

4.2.2   Feature Selection

So far RST has been successfully applied for feature selection. However, approaches have been proposed to use fuzzy-rough hybridization. In [4], the authors have presented a fuzzy-rough set based ant colony optimization algorithm and implemented it using fuzzy-rough hybridized approach to find the optimal feature subset.

Table 4.6 Fuzzy rough feature selection approaches

Fuzzy rough approach	Description
New approaches to fuzzy-rough feature selection [6]	Research proposes new approaches to fuzzy-rough FS-based on fuzzy similarity relations. A fuzzy extension to crisp discernibility matrices is presented as well
Different classes' ratio fuzzy rough set based robust feature selection [7]	Research proposes an effective robust fuzzy rough set model, called different classes' ratio fuzzy rough set (DC_ratio FRS) model to reduce the influence of noisy samples on the computation of the lower and upper approximations, and recognize the noisy samples directly
Large-Scale Multi-Modality Attribute Reduction with Multi-Kernel Fuzzy Rough Sets [9]	Research defines a combination of kernels based on set theory to extract fuzzy similarity for fuzzy classification with multi-modality attributes. Then, a model of multi-kernel fuzzy rough sets is constructed. Finally, the paper presents an attribute reduction algorithm for large-scale multi-modality fuzzy classification based on the proposed model
Towards scalable fuzzy–rough feature selection	Paper proposes two different novel ways to address this problem using a neighbourhood approximation step and attribute grouping in order to alleviate the processing overhead and reduce complexity
Fuzzy-rough feature selection accelerator	Paper proposes an accelerator, called forward approximation, which combines sample reduction and dimensionality reduction together. The strategy can be used to accelerate a heuristic process of fuzzy-rough feature selection. Based on the proposed accelerator, an improved algorithm is also designed
Semi-Supervised Fuzzy-Rough Feature Selection	Paper proposes a novel approach for semi-supervised fuzzy-rough feature selection where the object labels in the data may only be partially present. The approach also has the appealing property that any generated subsets are also valid (super)reducts when the whole dataset is labelled
Feature Selection Algorithm Using Fuzzy Rough Sets for Predicting Cervical Cancer Risks	Research uses fuzzy rough set method to analyze the demographic dataset and identify the risk of Cervical Cancer. This method integrates Entropy, Information Gain (IG) and Fuzzy Rough sets for identifying the risk of Cervical Cancer earlier. Risk Factors are identified by IG. Rules are extracted by Fuzzy Rough sets

The data in real world normally exists in hybrid format, so an effective technique for reduction of such data is also desirable. In [5], the authors propose an information measure to compute the discernibility power of a fuzzy or crisp equivalence relation and by using this measure the significance of different attributes was measured. Finally, the authors have also proposed two reduction algorithms for both supervised and unsupervised datasets.

Table 4.6 presents a few of the feature selection techniques based on fuzzy rough sets.

4.2.3 Rough Fuzzy Set

Dubios et al. in [6] defined the concept of Rough fuzzy sets to model the situation, where the knowledge base contains crisp concepts and output classes have poorly defined boundaries. It is rough set that is approximately deduced from fuzzy set in a crisp and approximate space. Output class is fuzzy in rough fuzzy set. For a rough fuzzy set FX, the lower and upper approximations are defined as follows.

Suppose FS = (U, A, V, f) is a knowledge representation system, if $P \subseteq A$ and $FX \subseteq U$ are fuzzy sets then the lower approximation $\underline{apr}_P(FX)$ and upper approximation $\overline{apr}_P(FX)$ of FS about FX is defined as

$$\underline{apr}_P(FX) = \inf\{x \in I\ (x) : \mu_{FX}(x)\} \tag{4.5}$$

$$\overline{apr}_P(FX) = \sup\{x \in I\ (x) : \mu_{FX}(x)\} \tag{4.6}$$

Where I is an equivalence relation on U and $\mu_{FX}(x)$ is the degree of membership of x to FX. The lower approximation defines the degree to which an object definitely belongs to fuzzy set FX, whereas upper approximation defines the degree of membership to which an object x possibly belongs to FX. So if for the values 0 or 1, these definitions are equal to those in conventional rough set theory.

If U = {X1, X2, X3, X4, X5} is a set of five employees and comprises of two equivalence classes U/I = {{X1, X3, X5}{X2, X4}}. Suppose a fuzzy set FX represents concept 'Competence' and membership function $\mu_{FX}(X) = \{\frac{X1}{0.5}, \frac{X2}{0.4}, \frac{X3}{0.1}, \frac{X4}{0.8}, \frac{X5}{0.7}\}$, the upper and lower approximation of the rough fuzzy set will be

$$\underline{apr}_P(FX) = \left\{\frac{X1}{0.1}, \frac{X2}{0.4}, \frac{X3}{0.1}, \frac{X4}{0.4}, \frac{X5}{0.1}\right\}$$

$$\overline{apr}_P(FX) = \left\{\frac{X1}{0.7}, \frac{X2}{0.8}, \frac{X3}{0.7}, \frac{X4}{0.8}, \frac{X5}{0.7}\right\}$$

4.2.4   Fuzzy Rough Set

Fuzzy rough sets are extensions of rough fuzzy set. According to Dubios in [6], the fuzzy rough set can be defined as follows.

The pair $\langle \underline{P}X, \bar{P}X \rangle$ is called a fuzzy rough set, where $\underline{P}X$ defines lower approximation and $\bar{P}X$ is called upper approximation as follows:

$$\mu_{\underline{P}X}(F_i) = \inf \, \max\{1 - \mu_{F_i}(x), \mu_X(x)\} \forall I \qquad (4.7)$$

$$\mu_{\bar{P}X}(F_i) = \inf \, \max\{\mu_{F_i}(x), \mu_X(x)\} \forall I \qquad (4.8)$$

Here
$F_i =$ *equivalence class*
$X =$ *fuzzy concept that requires approximation*
Obviously, when the equivalence relation is clear, a fuzzy rough set will degrade into a rough fuzzy set. When all the fuzzy equivalence relations are clear, it will further degrade into a classical rough set. So, the crisp lower approximation can be characterized by the following function [7]:

$$\mu_{\underline{P}X}(x) = \begin{cases} 1, & x \in F, \quad F \subseteq X \\ 0, & \text{otherwise} \end{cases} \qquad (4.9)$$

This means that if an object X belongs to equivalence class which is a subset of X, the object will belong to a lower approximation of X.

4.3   Dependency Classes

In conventional RST calculating the dependency of an attribute 'D' on 'C' requires scanning of the dataset and calculating the positive region which is a time-consuming job. In [8], the authors have presented a new concept of dependency classes. They developed an alternate way to calculate dependency comprising of dependency classes.

A dependency class is a heuristic which defines how the dependency measure changes as we scan new records during traversal of the dataset.

They start with the first record in dataset and calculate the dependency of decision attribute on conditional attribute based on the derived heuristics. Then after adding every single record the dependency of a particular attribute is refreshed based on to which decision class, the value of that attribute leads to. On the basis of the heuristics used by dependency classes, two types of dependency classes are proposed as follows:

- Incremental dependency classes [8],
- Direct dependency classes [9].

4.3.1 Incremental Dependency Classes (IDC)

Incremental dependency classes comprise four rules, where each rule defines a class that governs how dependency of decision attribute 'D' on 'C' changes as we read each new record.

We will explain each incremental dependency class with the help of an example. Consider the following decision system shown in Table 4.7 taken from [10]:

Here

$C = \{a', b', c', d'\}$

$D = \{D\}$ and $|U| = 10$

Initially, we start with the $|U| = 6$ where $U = \{a, b, d, e, f, g\}$. We calculate the dependency of 'D' on all attributes present in C (given at the end of each column in Table 4.8).

Now we define different classes through which the dependency can be calculated, after adding a new record.

Table 4.7 Decision system example

U	a'	b'	c'	d'	D
A	M	L	3	M	1
B	M	L	1	H	1
C	L	L	1	M	1
D	L	R	3	M	2
E	M	R	2	M	2
F	L	R	3	L	3
G	H	R	3	L	3
H	H	N	3	L	3
I	H	N	2	H	2
J	H	N	2	H	1

Table 4.8 Decision system example

U	a'	b'	c'	d'	D
A	M	L	3	M	1
B	M	L	1	H	1
D	L	R	3	M	2
E	M	R	2	M	2
F	L	R	3	L	3
G	H	R	3	L	3
	0.16667	0.3333	0.3333	0.5	

4.3.1.1   Existing Boundary Region Class

For an attribute a', if same value of attribute leads to different decision classes, for example, in Table 4.8, a'(L) -> 2,3 (i.e. the value 'L' leads to decision class '2' and '3') then adding a new record with same value of a' decreases the dependency of decision on that attribute. Adding a row in this case will simply decrease the dependency.

For example, in Table 4.8,

$$\gamma(a', D) = \frac{1}{6}$$

After adding new record, i.e. object 'C', the new dataset with new dependency values are shown in Table 4.9.

Before adding a new record: a'(L) -> 2,3 (in Table 4.8).

After adding a new record: a'(L) -> 1, 2, 3 (in Table 4.9).

So by adding a new record, the value 'L' of attribute 'a" which initially was leading to two decision classes, now leads to three decision classes, so γ(a', D) becomes

$$\gamma(a', D) = \frac{1}{7}$$

4.3.1.2   Positive Region Class

For an attribute a', if adding a record, does not lead to a different decision class for same value of that attribute, then dependency will increase. For example in Table 4.8, b'(L) -> 1. Previous dependency value was 2/6. After adding a new row (Object 'C' as shown in Table 4.9), b'(L) -> 1 sustains, i.e. the value 'L' of attribute b' uniquely identifies the decision class, so the new dependency will be

$$\gamma(b', D) = \frac{3}{7}$$

Table 4.9 adding new object 'C'

U	a'	b'	c'	d'	D
A	M	L	3	M	1
B	M	L	1	H	1
C	L	L	1	M	1
D	L	R	3	M	2
E	M	R	2	M	2
F	L	R	3	L	3
G	H	R	3	L	3
	0.14286	0.4286	0.4286	0.4286	

4.3.1.3   Initial Positive Region Class

For an attribute a′, if the value appears in the dataset for the first time for that attribute, then dependency increases. For example, adding a new record (object 'I') as shown in Table 4.10.

b′(N) -> 2. Initially, the value 'N' for b′ attribute was not present. Now adding the record for this value of b′ leads to new dependency value as follows.

$$\gamma(b, D) = \frac{4}{8}$$

4.3.1.4   Boundary Region Class

For an attribute a′, if the same value (which was leading to unique decision previously) of attribute leads to a different decision, then adding the new record reduces the dependency.

For example, adding a record 'H' in Table 4.11.

Table 4.10 adding new object 'I'

U	a′	b′	c′	d′	D
A	M	L	3	M	1
B	M	L	1	H	1
C	L	L	1	M	1
D	L	R	3	M	2
E	M	R	2	M	2
I	H	N	2	H	2
F	L	R	3	L	3
G	H	R	3	L	3
	0/8	0.5	0.5	0.25	

Table 4.11 adding new object 'H'

U	a′	b′	c′	d′	D
A	M	L	3	M	1
B	M	L	1	H	1
C	L	L	1	M	1
D	L	R	3	M	2
E	M	R	2	M	2
I	H	N	2	H	2
F	L	R	3	L	3
G	H	R	3	L	3
H	H	L	3	L	3
	0	1/9	0.5	0.25	

Table 4.12 Summary of IDC

Decision class	Definition	Initial attribute value	After adding new record	Effect on dependency
Existing Boundary region class	If the same value of attribute leads to different decision classes, it decreases the dependency	a′ (L) -> 2, 3	a′ (L) -> 1, 2, 3	Decreases
Positive region class	If adding a record, does not lead to a different decision class for the same value of that attribute, then dependency will increase	b′ (L) -> 1	b′ (L) -> 1	Increases
Initial Positive region class	If the value appears in the dataset for the first time for that attribute, then dependency increases	–	b′ (N) -> 2	Increases
Boundary region class	If the same value (which was leading to unique decision previously) of attribute leads to different decisions, then adding the new record reduces the dependency	b′ (L) -> 1	b′ (L) -> 1, 3	Decreases

The new dependency $\gamma(b', D)$ will be

$$\gamma(c, D) = \frac{1}{9}$$

Which was $\gamma(c, D) = \frac{3}{9}$ before adding record 'H'. Table 4.12 shows the summary of all decision classes.

4.3.1.5   Mathematical Representation of IDC

Now we provide a mathematical representation of IDC and an example about how to calculate IDC. Mathematically:

$$\gamma(Attribute, D) = \frac{1}{N} \sum_{k=1}^{N} \gamma'_k \tag{4.10}$$

Where:

$$\gamma'_k = \begin{cases} 1, & if\ V_{Attribute,k}\ leades\ to\ a\ positive\ region\ class \\ 1, & if\ V_{Attribute,k}\ leades\ to\ an\ initial\ positive\ region\ class \\ -n, & if\ V_{Attribute,k}\ leades\ to\ a\ boundary\ region\ class \\ 0, & x\ if\ V_{Attribute,k}\ has\ already\ lead\ to\ boundary\ region\ class \end{cases}$$

Table 4.13 Symbols and their semantics

Symbol	Semantics
$\gamma\,(Attribute, D)$	Dependency of attribute 'D' on attribute '*Attribute*'
'*Attribute*'	Name of current attribute under consideration
D	Decision attribute (Decision Class)
γ'_k	Dependency value contributed by object 'k' for attribute '*Attribute*'
$V_{Attribute,\ k}$	Value of attribute '*Attribute*' for object 'k' in the dataset
n	Total number of previous occurrences of $V_{Attribute,\ k}$
N	Total number of records in dataset

Table 4.13 represents the symbol and its semantics.

4.3.1.6 Example

Following example shows how IDC calculates dependency. We read each record and identify its dependency class. Based on the class we decide the factor by which dependency will be added to the overall dependency value. We will consider the dataset given in Table 4.7. Using IDC:

$$\gamma\,(Attribute, D) = \frac{1}{N}\sum_{k=1}^{N}\gamma'_k$$

Consider the attribute {a'}, we read first three records, i.e. object 'A', as it appears for the first time, so it belongs to 'Initial Positive Region' class, thus we will add '1' to overall dependency value. Reading objects 'B' and 'C' lead to 'Positive region' class, so we will add '1' for both. Reading 'D', the value 'L' now leads to decision class '2' (previously its one occurrence was leading to decision class '1'), so it belongs to 'Boundary region' class and thus we will add the value '−1' to overall dependency. Reading object 'E' at this stage, value 'M' belongs to 'Boundary region' class and it had two occurrences before, so, we will add '−2' in overall dependency. Reading object 'F', note that it has already lead to 'Boundary region' class, so we will add '0' to overall dependency and so on.

$$\gamma(\{a'\}, D) = \frac{1}{10}\sum_{k=1}^{10}\gamma'_k = (1+1+1+(-1)+(-2)+0+1+1+(-2)+0) = \frac{0}{10}$$
$$= 0$$

Similarly, dependency of 'D' on {b', c', d'} will be as follows:

$$\gamma(\{b'\}, D) = \frac{1}{10}\sum_{k=1}^{10}\gamma'_k = (1+1+1+1+1+(-2)+0+1+(-1)+0) = \frac{3}{10}$$

$$\gamma(\{c'\}, D) = \frac{1}{10}\sum_{k=1}^{10}\gamma'_k = (1+1+1+(-1)+1+0+0+0+1+(-2)) = \frac{2}{10}$$

$$\gamma(\{d'\}, D) = \frac{1}{10}\sum_{k=1}^{10}\gamma'_k = (1+1+1+(-2)+0+1+1+1+(-1)+0) = \frac{3}{10}$$

Note that if a value of an attribute has already lead to boundary region class than adding the same value will simply be reflected by adding '0' as dependency.

4.3.2 Direct Dependency Classes (DDC)

Direct dependency classes are alternate to IDC for calculating dependency directly without involving positive region and exhibit almost the same performance as shown by IDC. DDC determines the number of unique/non-unique classes in a dataset for an attribute C. *A unique class represents the attribute values that lead to unique decision class throughout dataset, so this value can be used to precisely define decision class.*

For example in Table 4.8, the value 'L' of attribute b' is unique class as all of its occurrences in the same table lead to a single/unique decision class (i.e. '1'). *Non-unique classes represent the attribute values that lead to different decision classes, so they cannot be precisely used to determine the decision class.* For example in Table 4.8, the value 'R' of attribute b' represents the non-unique class as some of its occurrences lead to decision class '2' and some occurrences lead to decision class '3'.

Calculating unique/non-unique classes directly lets us avoid complex compu-tations of the positive region. The basic idea behind the proposed approach is that the number of unique classes increase dependency and non-unique classes decrease dependency. For a decision class D, the dependency K of D on C is shown in Table 4.14.

The dependency using DDC approach can be calculated by the following formula:

Table 4.14 How DDC calculates dependency

Dependency	No. of unique/non-unique classes
0	If there is no unique class
1	If there is no non-unique class
n	Otherwise where $0 < n < 1$

For unique dependency classes:

$$\gamma(Attribute, D) = \frac{1}{N}\sum_{i=1}^{N}(\gamma_i')$$ (4.11)

$$\gamma_i' = \begin{cases} 1, & \text{if classisunique} \\ 0, & \text{if classisnon} - \text{unique} \end{cases}$$

And for non-unique dependency classes:

$$\gamma(Attribute, D) = 1 - \frac{1}{N}\sum_{i=1}^{N}(\gamma_i')$$ (4.12)

$$\gamma_i' = \begin{cases} 0, & \text{if classisunique} \\ 1, & \text{if classisnon} - \text{unique} \end{cases}$$

Table 4.15 represents the symbol and its semantics.

4.3.2.1 Example

We consider the decision system given in Table 4.7. As per definitions of unique dependency classes:

$$\gamma(Attribute, D) = \frac{1}{N}\sum_{i=1}^{N}(\gamma_i')$$

If consider attribute $\{b'\}$, there are three unique dependency classes, i.e. there are three occurrences of value 'L' that lead to unique decision class, so

$$\gamma(\{b'\}, D) = \frac{1}{10}\sum_{i=1}^{10}(\gamma_i')$$

Table 4.15 Symbols and their semantics

Symbol	Semantics
γ (Attribute, D)	Dependency of attribute 'D' on attribute 'Attribute'
'Attribute'	Name of current attribute under consideration
D	Decision attribute (Decision Class)
γ_i'	Dependency value contributed by object 'I' for attribute 'Attribute'
N	Total number of records in dataset

$$\gamma(\{b'\}, D) = \frac{1}{10}(1+1+1+0+0+0+0+0+0+0) = \frac{3}{10}$$

Similarly for attribute $\{c'\}$:

$$\gamma(\{c'\}, D) = \frac{1}{10}\sum_{i=1}^{10}(\gamma'_i)$$

$$\gamma(\{c'\}, D) = \frac{1}{10}(0+1+1+0+0+0+0+0+0+0) = \frac{2}{10}$$

On the other hand, if we consider non-unique dependency classes:

$$\gamma(Attribute, D) = 1 - \frac{1}{N}\sum_{i=1}^{N}(\gamma'_i)$$

If we consider attribute 'b'', note that there are seven non-unique dependency classes (four occurrences of value 'R' lead to two decision classes and three occurrences of value 'N' lead to three decision classes), so

$$\gamma\left(\{b'\}, D\right) = 1 - \frac{1}{10}\sum_{i=1}^{10}(\gamma'_i) = 1 - \frac{1}{10}(0+0+0+1+1+1+1+1+1+1)$$

$$= \frac{3}{10}$$

Similarly

$$\gamma(\{c'\}, D) = 1 - \frac{1}{10}\sum_{i=1}^{10}(\gamma'_i) = 1 - \frac{1}{10}(1+0+0+1+1+1+1+1+1+1)$$

$$= \frac{2}{10}$$

Note that there are three unique decision classes in attribute $\{b'\}$ and seven in $\{c'\}$. For a decision system:

No of unique classes + no of nonunique classes = size of universe

So we either need to calculate the number of unique classes or non-unique classes. The algorithm for DDC is shown in Fig. 4.7.

Grid is the main data structure used to calculate dependency directly without using a positive region. It is a matrix with the following dimensions:

No. of rows = No. of records in the dataset.

No. of columns = number of conditional attributes + number of decision attributes + 2.

So if there are ten records in dataset, five conditional attributes and one decision class then grid dimension will be 10 × 8, i.e. ten rows and eight columns. A row read from the dataset will first be stored in the grid if it does not already exist. The five conditional attributes will be stored in the first five columns; decision attribute will be stored in sixth column called DECISIONCLASS. Seventh column called INSTANCECOUNT will store the number of times that record appears in the actual dataset, finally, the last column called CLASSSTATUS will store the uniqueness of the record, the value '0' means record is unique and '1' means it is non-unique. If a record, read from dataset already exists in Grid, its INSTANCECOUNT will be incremented. If the decision class of the record is different from that already stored in Grid, i.e. the same values of attributes now lead to different decision class, CLASSSTATUS will be set to '1'. However, if the record is inserted for the first time, INSTANCECOUNT is set to '1' and CLASSSTATUS is set to 0, i.e. it is considered unique.

The functions 'FindNonUniqueDependency' and 'FindUniqueDependency' are main functions to calculate the dependency. Functions insert the first record in Grid and then search for the same record in the entire dataset. The status of the record is updated in Grid as soon as further occurrences of the same record are found. Finally, the functions calculate the dependency value on the basis of uniqueness/ non-uniqueness of classes. Function 'InsertInGrid' inserts the record in the next row of the Grid. 'FindIndex' finds the row no. of the current record in the Gird. 'IfAlreadyExistsInGrid' finds if the record already exists or not. Finally, 'UpdateUniquenessStaus' function updates the status of the record in Grid.

4.4 Redefined Approximations

Unique decision classes lead to the idea of calculating lower and upper approximation without using indiscernibility relation [11]. Calculating lower and upper approximation requires calculating equivalence classes (indiscernibility relation) which is computationally an expensive task. Using unique decision classes lets us avoid this task and we can directly calculate lower approximation. The new definitions are semantically the same to the conventional definitions but provide a computationally more efficient method for calculating these approximations by avoiding equivalence class structure. The following section discusses the proposed new definitions in detail.

4.4.1 Redefined Lower Approximation

The conventional rough set based lower approximation defines the set of objects that can with certainty be classified as members of concept 'X'. For attribute(s) $c \in C$ and concept X, the lower approximation will be

```
Function FindNonUniqueDependency                    Function FindUniqueDependency
Begin                                               Begin
InsertInGrid(X₁)                                     InsertInGrid(X₁)
For i=2 to TotalUnivesieSize                         For i=2 to TotalUnivesieSize
IfAlreadyExistsInGrid(Xᵢ)                            IfAlreadyExistsInGrid(Xᵢ)
     Index = FindIndexInGrid(Xᵢ)                          Index = FindIndexInGrid(Xᵢ)
     If DecisionClassMatched(index,i) = False            If DecisionClassMatched(index,i) = True
UpdateUniquenessStaus(index)                         UpdateUniquenessStaus(index)
     End-IF                                               End-IF
  Else                                                  Else
InsertInGrid(Xᵢ)                                     InsertInGrid(Xᵢ)
  End-IF                                                End-IF

Dep=0                                               Dep=0

For i=1 to TotalRecordsInGrid                        For i=1 to TotalRecordsInGrid
 If Grid(I,CLASSSTATUS) = 1                           If Grid(i,CLASSSTATUS) = 0
   Dep= Dep+ Grid(i,INSTANCECOUNT)                      Dep= Dep+ Grid(i,INSTANCECOUNT)
End-IF                                               End-IF
Return (1-Dep)/TotalRecords                          Return Dep/TotalRecords
End Function                                         End Function

Function InsertInGrid(Xi)                            Function InsertInGrid(Xi)
For j=1 to TotalAttributesInX                        For j=1 to TotalAttributesInX
 Grid(NextRow,j) = Xᵢʲ                                Grid(NextRow,j) = Xᵢʲ
End-For                                              End-For
 Grid(NextRow,DECISIONCLASS) = Dᵢ                     Grid(NextRow,DECISIONCLASS) = Dᵢ
Grid(NextRow, INSTANCECOUNT) = 1                     Grid(NextRow, INSTANCECOUNT) = 1
Grid(NextRow, CLASSSTATUS) = 1 // 1 => uniqueness    Grid(NextRow, CLASSSTATUS) = 1 // 1 => uniqueness
End-Function                                         End-Function

Function IfAlreadyExistsInGrid(Xᵢ)                   Function IfAlreadyExistsInGrid(Xᵢ)
 For i=1 to TotalRecordsInGrid                        For i=1 to TotalRecordsInGrid
   For j=1 to TotalAttributesInX                       For j=1 to TotalAttributesInX
     If Grid(i,j) <>Xʲ                                  If Grid(i,j) <>Xʲ
       Return False                                       Return False
   End-For                                             End-For
 End-For                                              End-For
Return True                                          Return True
End-Function                                         End-Function

Function FindIndexInGrid(Xᵢ)                         Function FindIndexInGrid(Xᵢ)
 For i=1 to TotalRecordsInGrid                        For i=1 to TotalRecordsInGrid
RecordMatched=TRUE                                   RecordMatched=TRUE
   For j=1 to TotalAttributesInX                       For j=1 to TotalAttributesInX
   If Grid(i,j) <>Xʲ                                   If Grid(i,j) <>Xʲ
RecordMatched=FALSE                                  RecordMatched=FALSE
   End-For                                             End-For
   If RecordMatched= TRUE                              If RecordMatched= TRUE
     Return j                                            Return j
   End-If                                              End-If
 End-For                                             End-For
Return True                                          Return True
End-Function                                         End-Function

Function DecisionClassMatched(index,i)              Function DecisionClassMatched(index,i)
 If Grid(index, DECISIONCLASS) = Dᵢ                  If Grid(index, DECISIONCLASS) = Dᵢ
   Return TRUE                                         Return TRUE
 Else                                                Else
   Return False                                        Return False
 End-If                                              End-If
End-Function                                         End-Function

Function UpdateUniquenessStaus(index)               Function UpdateUniquenessStaus(index)
Grid(index, CLASSSTATUS) = 1                         Grid(index, CLASSSTATUS) = 1
End-Function                                         End-Function
```

DDC using non-unique classes	DDC using Unique Classes

Fig. 4.7 Pseudo code for directly finding dependency using unique and non-unique classes

$$\underline{C}X = \{X|[X]_c \subseteq X\} \tag{4.13}$$

This definition requires the calculation of indiscernibility relation, i.e. equivalence class structure $[X]_c$, where the objects belonging to one equivalence class are indiscernible with respect to the information present in attribute(s) $c \in C$. Based on the concept of lower approximation provided by RST, we have proposed a new definition as follows:

$$\underline{C}X = \{\forall x \in U, c \in C, a \neq b | x_{\{c \cup d\}} \rightarrow a, x_{\{c \cup d\}} \nrightarrow b\} \tag{4.14}$$

i.e. the lower approximation of concept 'X' with respect to the attribute set 'c' is set of objects such that for each occurrence of the object, the same value of conditional attribute 'c' always leads to the same decision class value. So, if there are 'n' occurrences of an object, then all of them lead to same decision class (for the same value of attributes), which alternatively means that for a specific value of an attribute, we can with surety say that object belongs to a certain decision class. This is exactly equal to the conventional definition of lower approximation.

So, the proposed definition is semantically the same as conventional one, however, computationally it is more convenient for calculating lower approximation, it avoids construction of equivalence class structures to find the objects belonging to the positive region. It directly scans the dataset and finds the objects that lead to the same decision class throughout thus enhancing the performance of the algorithm using this measure. We will use Table 4.16 as a sample to calculate lower approximation using both definitions.

We suppose the concept: $X = \{x|Z(x) = 2\}$, i.e. we will find the objects about which could with surety say that they lead to decision class '2'. The conventional definition requires three steps given below:

Step-1: Calculate the objects belonging to the concept X. Here is our case concept $X = \{x|Z(x) = 2\}$

So, we will identify the objects belonging to the concept. In our case

Table 4.16 Sample decision system

U	a	b	c	d	Z
X1	L	3	M	H	1
X2	M	1	H	M	1
X3	M	1	M	M	1
X4	H	3	M	M	2
X5	M	2	M	H	2
X6	L	2	H	L	2
X7	L	3	L	H	3
X8	L	3	L	L	3
X9	M	3	L	M	3
X10	L	2	H	H	2

$$X = \{X4, X5, X6, X10\}$$

Step-2: Calculating equivalence classes using conditional attributes.

Here we will consider only one attribute 'b' for simplicity, i.e. $C = \{b\}$. So, we will calculate equivalence classes using attribute 'b', which in our case becomes

$$[X]_c = \{X1, X4, X7, X8, X9\}\{X2, X3\}\{X5, X6, X10\}$$

Step-3: Find objects belonging to lower approximation.

In this step, we actually find the objects that belong to lower approximation of the concept w.r.t. to considered attribute. Mathematically, this step involves finding objects from $[X]_c$ which are a subset of X, i.e. $\{[X]_c \subseteq X\}$. In our case:

$$\underline{C}X = \{[X]_c \subseteq X\} = \{X5, X6, X10\}$$

Using the proposed definition, we construct the lower approximation without using equivalence class structures. We directly find the objects that under the given concept always lead to the same decision class (concept) for the current value of attributes.

In our case, we just pick an object and find if, for the same values of attributes, it leads to some other decision class or not. We find that objects X5, X6 and X10 always lead to the same decision class, i.e. the concept under consideration for attribute 'b'. On the other hand objects, X2 and X3 do not lead to Concept X, so they will not be part of lower approximation. Similarly some occurrences of objects X1, X4, X7, X8 and X9 also lead to different decision class than X, so they also be excluded from lower approximation. So, we can with surety say that objects X5, X6 and X10 belong to the lower approximation.

As discussed earlier, semantically both definitions are the same but computationally the proposed definition is more effective as it avoids construction of equivalence class structure. It directly calculates the objects belonging to lower approximation by checking objects that always lead to the same decision class under consideration. Directly calculating the lower approximation in this way lets us exclude complex equivalence class structure calculation which makes algorithms using this measure more efficient. Using conventional definition, on the other hand, requires three steps discussed in the previous section. Performing these steps offers a significant performance bottleneck to algorithms using this measure.

4.4.2   Redefined Upper Approximation

Upper approximation defines the set of objects that can only be classified as possible members of X with respect to the information contained in attribute(s) $c \in C$. For attribute(s) $c \in C$ and concept X, the lower approximation will be

$$\bar{C}X = \{X | [X]_B \cap X \neq 0\}$$

i.e. upper approximation is the set of objects that may belong to the concept X with respect to information contained in attribute(s) c.

On the bases of the same concept, a new definition of upper approximation was proposed as follows:

$$\bar{C}X = \underline{C}X \cup \left\{ \forall x \in U, c \in C, a \neq b | x_{\{c\}} \rightarrow a, x'_{\{c\}} \rightarrow b \right\}$$

This definition will be read as follows.

Provided that that objects x and x' are indiscernible with respect to attribute(s) c, they will be part of an upper approximation if either they belong to lower approximation or at least one of their occurrences leads to decision class belonging to concept X. So objects x and x' belong to upper approximation if both occurrences of them lead to different decision class for the same value of attributes. For example, in Table 4.16 the objects X1, X4, X7, X8, and X9 lead to different decision class for the same value of attribute 'b'. So all of them can be possible members of upper approximation of concept Z = 2.

As with redefined lower approximation, the proposed definition for the upper approximation is semantically the same as a conventional upper approximation. However, it helps us directly calculate objects belonging to the upper approximation without calculating the underlying equivalence class structures.

Like the conventional definition of lower approximation, calculating upper approximation requires three steps again. We will again use Table 4.16 as a sample to calculate upper approximation using both definitions. We suppose the concept: $X = \{x | Z(x) = 2\}$ as shown in the example of lower approximation.

Step-1: Calculate the objects belonging to the concept X. Here is our case concept $X = \{x | Z(x) = 2\}$.

We have already performed this step and calculated the objects belonging the concept X on the basis of decision class in previous example. So,

$$X = \{X4, X5, X6, X10\}$$

Step-2: Calculating equivalence classes using conditional attributes.

This step has also been calculated before and objects belonging to $[X]_c$ were identified as follows by considering attribute 'b':

$$[X]_c = \{X1, X4, X7, X8, X9\}\{X2, X3\}\{X5, X6, X10\}$$

Step-3: find the objects belonging to upper approximation.

In this step, we finally calculate the objects belonging to lower approximation. The step is calculated by identifying the objects in $[X]_c$ that have non-empty interaction with objects belonging to concept X. The intention is to identify all those objects at least one instance of which leads to the decision class belonging to the concept X. Mathematically $[X]_c \cap X \neq 0$. In our case:

$$\bar{C}X = \{X1, X4, X7, X8, X9\}\{X5, X6, X10\}$$

The proposed definition of upper approximation avoids calculating computational expensive step of indiscernibility relation. It simply scans the entire dataset for the concept X using the attributes c and finds all the indiscernible objects such that any of their occurrences belongs to the concept X. At least one occurrence should lead to decision class belonging to concept X. In our case, objects X1, X4, X7, X8 and X9 are indiscernible and at least one of their occurrence leads to concept Z = 2. Objects X2 and X3 are indiscernible but none of their occurrences leads to the required concept, so they will not be part of upper approximation. Objects X5, X6 and X10 belong to lower approximation so they will be part of upper approximation as well. In this way using the proposed definition, we find the following objects as part of the upper approximation:

$$\bar{C}X = \{X1, X4, X7, X8, X9\}\{X5, X6, X10\}$$

That is the same as produced by using the conventional method.

The semantically proposed definition of upper approximation is the same as the conventional one, i.e. it produces the same objects that may possibly be classified as members of concept X. However, it calculates these objects without calculating equivalence classes. So, the proposed approach is computationally less expensive which consequently can result in enhanced performance of the algorithms using the proposed method. The conventional definition, on the other hand, again requires three steps (as discussed above) to calculate upper approximation which imposes computational implications.

When a problem in real world becomes so complex that you cannot decide a problem to be true or false or you cannot predict the accuracy of a statement to be one or zero, in that case, we say its probability is uncertain. Normally, we solve the problems through probability but when u get a system where all elements are vague and not clear then we use fuzzy theory to handle such systems. Fuzzy means vague, something that is not clear.

Now what is the difference between fuzzy and crisp set? We can explain with a simple argument. If something can be answered with yes or no then it means it is crisp, e.g. is water colorless? Definitely answer will be yes, so this statement is crisp, but what about the statement is John honest? Now you cannot precisely answer this question because someone would think that john is more honest, yet some other would take him less honest, etc. So, this statement will not be crisp but fuzzy.

Now the question is that why does fuzzy situation arise? It arises due to our partial information about the problem or due to information that is not fully reliable or inherent imprecision in the statement due to which we will say that the problem is fuzzy.

4.5 Summary

In this chapter, we have presented some advance concepts of rough set theory, we discussed advanced forms of three fuzzy concepts, i.e. lower approximation, upper approximation and dependency. Then some preliminary concepts regarding fuzzy set theory, hybridization of RST with fuzzy set theory was presented.

References

1. Lingras P, Jensen R (2007) Survey of rough and fuzzy hybridization. In: 2007 IEEE International Fuzzy Systems Conference. IEEE
2. Sarkar M (2000) Fuzzy-rough nearest neighbors algorithm. In: 2000 IEEE International Conference on Systems, Man, and Cybernetics, vol 5. IEEE
3. De Cock M, Cornelis C (2005) Fuzzy rough set based web query expansion. In: Proceedings of Rough Sets and Soft Computing in Intelligent Agent and Web Technology, International Workshop at WIIAT2005
4. Jensen Richard, Shen Qiang (2005) Fuzzy-rough data reduction with ant colony optimization. Fuzzy Sets Syst 149(1):5–20
5. Hu Qinghua, Daren Yu, Xie Zongxia (2006) Information-preserving hybrid data reduction based on fuzzy-rough techniques. Pattern Recogn Lett 27(5):414–423
6. Dubois Didier, Prade Henri (1990) Rough fuzzy sets and fuzzy rough sets. International Journal of General System 17(2–3):191–209
7. Jensen R, Shen Q (2008) Computational intelligence and feature selection: rough and fuzzy approaches, vol 8. Wiley
8. Raza MS, Qamar U (2016) An incremental dependency calculation technique for feature selection using rough sets. Inf Sci 343:41–65
9. Raza MS, Qamar U (2018) Feature selection using rough set-based direct dependency calculation by avoiding the positive region. Int J Approx Reason 92:175–197
10. Al Daoud E (2015) An efficient algorithm for finding a fuzzy rough set reduct using an improved harmony search. Int J Mod Educ Comput Sci 7(2):16
11. Raza MS, Qamar U (2016) Feature selection using rough set based dependency calculation. Phd Dissertation, National University of Science and Technology

Chapter 5
Rough Set Theory Based Feature Selection Techniques

Rough Set Theory has been successfully used for feature selection techniques. The underlying concepts provided by RST help in finding representative features by eliminating the redundant ones. In this chapter, we will present various feature selection techniques which use RST concepts.

5.1 Quick Reduct

In QuickReduct (QR) [1], the authors attempt to develop a forward feature selection mechanism without exhaustively generating all possible subsets. The algorithm starts with an empty set and adds attributes one by one, which result in a maximum increase in the degree of dependency. The algorithm continues until maximum dependency value is achieved. After adding each attribute, the dependency is calculated and the attribute is kept if it results in a maximum increase in dependency. If at any stage, the value of the selected attribute set becomes equal to that of the entire dataset algorithm terminates with currently selected subset as Reduct. Figure 5.1 shows the pseudocode of the algorithm.

The algorithm uses positive-region-based approach for calculating the dependency at steps 5 and 8. However, using positive-region-based dependency measure requires three steps, i.e., calculating equivalence classes using decision attribute, calculating dependency classes using conditional attributes, and finally calculating positive region. Using these three steps can be computationally expensive.

Since QR is one of the most common RST-based feature selection algorithms, we explain it with the help of an example, but first, we see how does it work? The algorithm starts with an empty Reduct Set R, it then uses a loop to calculate the dependency of each attribute subset. Initially, the subset comprises a single attribute. The attribute subset with maximum dependency is found at the end of each iteration and is considered as Reduct Set R. For next iteration, previous attribute

© Springer Nature Singapore Pte Ltd. 2019
M. S. Raza and U. Qamar, *Understanding and Using Rough Set
Based Feature Selection: Concepts, Techniques and Applications*,
https://doi.org/10.1007/978-981-32-9166-9_5

Fig. 5.1 QuickReduct
algorithm taken from [1]

QUICKREDUCT(C, D)
C, The set of all conditional features
D, The set of decision features
    (1) $R \leftarrow \{\}$
    (2) do
    (3) $T \leftarrow R$
    (4) $\forall x \in (C - R)$
    (5) $If \gamma_{R \cup \{x\}}(D) > \gamma_T(D)$
    (6) $T \leftarrow R \cup \{x\}$
    (7) $R \leftarrow T$
    (8) $until \gamma_R(D) == \gamma_c(D)$
    (9) Output R

subset with maximum dependency is joined with every single attribute in conditional attribute set and a subset with maximum dependency is found again. The process continues until at any stage, an attribute subset with dependency equal to that of entire conditional attribute set is found. The sequence is as follows:

(1) Find the dependency of each attribute.
(2) Find attribute with maximum dependency and make it R.
(3) Start the next iteration.
(4) Join R with every single attribute in {C}–{R}.
(5) Again calculate the subset with maximum dependency.
(6) Now R comprises two attributes.
(7) Again join R with every single attribute in {C}–{R}.
(8) Again calculate subset with maximum dependency.
(9) Now R comprises three attributes.
(10) Repeat the entire process until at any stage, the dependency of R = 1 or that of the entire set of conditional attributes.

We now explain this with the help of an example. We will consider the following dataset (Table 5.1).

Table 5.1 Sample decision system

U	a	b	c	d	Z
X1	L	3	M	H	1
X2	M	1	H	M	1
X3	M	1	M	M	1
X4	H	3	M	M	2
X5	M	2	M	H	2
X6	L	2	H	L	2
X7	L	3	L	H	3
X8	L	3	L	L	3
X9	M	3	L	M	3
X10	L	2	H	H	2

Here, $C = \{a, b, c, d\}$ are conditional attributes whereas $D = \{Z\}$ is decision class.

Initially $R = \{\phi\}$. Join R with each attribute in the conditional set and find the attribute subset with maximum dependency:

$$\gamma(\{R \cup A\}, Z) = 0.1$$

$$\gamma(\{R \cup B\}, Z) = 0.5$$

$$\gamma(\{R \cup C\}, Z) = 0.3$$

$$\gamma(\{R \cup D\}, Z) = 0.0$$

So far $\{R \cup B\}$ results in a maximum increase in dependency, hence $R = \{b\}$. Now, we join R with each conditional attribute except $\{b\}$.

$$\gamma(\{R \cup A\}, Z) = 0.6$$

$$\gamma(\{R \cup C\}, Z) = 0.8$$

$$\gamma(\{R \cup D\}, Z) = 0.6$$

So, $\{R \cup C\}$ results in a maximum increase in dependency, hence $R = \{b, c\}$. Now, again we join R with every conditional attribute except $\{b, c\}$.

$$\gamma(\{R \cup A\}, Z) = 1.0$$

$$\gamma(\{R \cup D\}, Z) = 1.0$$

Note that both $\{R \cup A\}$ and $\{R \cup D\}$ result in the same degree of increase in dependency, so the algorithm will pick the first one, i.e., $\{R \cup A\}$ and will output $R = \{a, b, c\}$ as Reduct. Here, we find a very interesting aspect of the algorithm that the performance of this algorithm depends on the distribution of the conditional attributes as well. There are many variations proposed on QR in intention to optimize the feature selection process. One is REVERSEREDUCT [1] and other is accelerated QR [2].

REVERSEREDUCT [1] is the strategy for attribute reduction, however, it uses backward elimination in contrast with forward feature selection mechanism. The algorithm starts by considering the entire set of conditional attributes as Reduct. It then removes one attribute at a time and calculates dependency until the removal of any further attribute becomes impossible without introducing inconsistency. The algorithm also suffers the same problem as that faced by QuickReduct. It uses positive-region-based dependency measure and is equally unsuitable for larger datasets.

In accelerated QR, on the other hand, attributes are selected in order of degree of dependency. The algorithm starts by calculating the dependency of each conditional

Accelerated Quickreduct(C, D)
C: c_1, c_2, ..., c_n, the set of all conditional features;
D: d, a decision features
    (a) R ←{}
    (b) $\gamma_{prev} = 0$, $\gamma_{best} = 0$
    (c) Do
    (d) T ← R
    (e) $\gamma_{prev} = \gamma_{best}$
    (f) *Compute* γ_i,   i   =   1...n   where $\gamma_R(D)$ = card($POS_R(D)$)/card(U)
                     $POS_R(D) = \_RX$
    (g) Select Max(γ_i) *or* (γ_j) where $j \in$ { the set of all attributes which have the same highest degree of dependency}
    (h) if $\gamma_{RUX}(D) > \gamma_{prev}(D)$
    (i) T ← RU{X}
    (j) $\gamma_{best} = \gamma_T(D)$
    (k) R ← T
    (l) until $\gamma_{best} == \gamma_{prev}$
    (m) return R

Fig. 5.2 Accelerated QR [2]

attribute and selects the attribute with maximum dependency. If two or more attributes have the same degree of dependency then both are combined and dependency is calculated again. This value is then compared with the previous value. If the condition becomes 'false', the attributes are chosen in combination with the next highest degree of dependency value and the process is repeated until the condition becomes 'true'. Figure 5.2 shows the pseudocode of the accelerated QR algorithm.

5.2 Hybrid Feature Selection Algorithm Based on Particle Swarm Optimization (PSO)

In [3], Hanna et al. presented a supervised hybrid feature selection algorithm based on Particle Swarm Optimization (PSO) and RST. The algorithm computes Reducts without exhaustively generating all possible subsets. The algorithm starts with an empty set and adds attributes one by one. It constructs a population of particles with random position and velocity in S dimensions. In the problem space, it then computes the fitness function of each particle using RST-based dependency measure. The feature with the highest dependency is selected and the combination of all other features with this one is constructed. Fitness of each of these combinations is selected. If the fitness value of this particle is better than the previous best (pbest) value, this is selected as pbest. Its position and fitness are stored. It then compares the fitness of current particle with the population's overall previous best fitness (gbest). If it is better than gbest, then gbest position is set to current the current particle's position with the global best fitness updated. This position represents the best feature subset encountered so far, and is stored in R. The algorithm then updates the velocity and the position of each particle. It continues until the stopping criteria is met, which is the maximum number of iterations in normal case. According to the algorithm, the dependency of each attribute subset is calculated based on the dependency on decision attribute and the best particle is chosen. The algorithm uses positive-region-based dependency measure and is the enhancement of QuickReduct algorithm.

The velocity of each particle is represented using a positive number from 1 to V_{max}. It implies that how many bits of a particle should be changed to be the same as that of the global best position. The difference in the number of bits between two particles implies the difference between their positions, e.g., if Pbest = [1, 0, 1, 1, 1, 0, 1, 0, 0, 1], Xi = [0, 1, 0, 0, 1, 1, 0, 1, 0, 1] then the difference between Pgbest and Xi is Pgbest − Xi = [1, −1, 1, 1, 0, −1, 1, −1, 0, 0]. '1' means that this bit (each '1' represents the presence of feature and '0' represents absence) should be selected as compared to the global best position and '−1' means that this bit should not be selected. After velocity is updated, the next task is to update the position by new velocity. If the new velocity is V, and the number of different bits between the current particle and gbest is xg, then the position is updated as per the following conditions:

- V ≤ xg. In this case, random V bits, which are different from that of gbest, are changed. So the particle will move toward the best position while keeping its exploration ability.
- V > xg. In this case, apart from the bits to be the same as that of gbest, (V − xg) further bits should also be randomly changed. Hence, after the particle reaches the global best position, it keeps on moving some distance toward other directions, which gives it further exploration ability.

Figure 5.3 shows the pseudocode of PSO-QR algorithm.

5.3 Genetic Algorithm

In [4], the authors present a rough set based Genetic Algorithm (GA) for feature selection. The selected set of features was provided to artificial neural network classifier for further analysis. The algorithm uses positive-region-based dependency measure as fitness for generated candidates in the proposed system. The proposed

Input: C, the set of all conditional features,
            D, the set of decision features.
Output: Reduct R

Step 1: Initialize X with random position V_i with random velocity
$\forall: X_i \leftarrow randomPosition()$;
$V_i \leftarrow randomVelocity()$;
Fit $\leftarrow$ 0; globalbest $\leftarrow$ Fit;
Gbest $\leftarrow$ X_1; Pbest(1) $\leftarrow$ X_1
For i = 1 ... S
pbest(i) = X_i
Fitness(i) = 0
End For

Step 2: While Fit != 1   // stopping criterion
For i = 1 ... S // for each particle
$\forall: X_i$;
//Compute fitness of feature subset of X_i
R $\leftarrow$ Feature subset of X_i (1's of X_i)
$\forall x \in (C - R)$
$$\gamma_{RU(X)}(D) = \frac{\left|POS_{RU(X)}(D)\right|}{|U|}$$

$Fit = \gamma_{RU(X)}(D) \ \forall \ xcR, \gamma_x(D) \neq \gamma_c(D)$

End For

Step 3: Compute best fitness
For i = 1:S
if(Fitness(i) > globalbest) // if current fitness is greater than global best fitness
globalbest $\leftarrow$ Fitness(i); // assign current fitness value as global best fitness
gbest $\leftarrow$ X_i ;
getReduct(X_i)
Exit
End if
End for
UpdateVelocity(); // Update velocity V_i's of X_i's
UpdatePosition(); // Update position of X_i's
// Continue with the next iteration
End {while}
Output Reduct R

Fig. 5.3 PSO-QR taken from [3]

system uses RST-based feature dependency value of each chromosome for finding the high-performance optimal Reducts. Stopping criterion was defined on the based on the following equation:

$$k = \gamma(C, D) = \frac{|POS_C(D)|}{|U|} \geq \alpha.$$

The candidates equal or greater than were accepted as a result. The following equation was used to calculate the solution addition type added to the solution:

$$RSC\% = 100\% - (BSC\% + WSC\%)$$

where

RSC = Random Selected Chromosomes,
BSC = Best Solution Candidates,
WSC = Worst Solution Candidates.

The number of generations in each generation pool was 2*n, where 'n' is user-defined parameter and can be changed by the user for optimal performance and specifying the number of generations. In the proposed version, these 2*n (2, 4, 6... n) generations were randomly initialized and used for generating the following generations. The last 2*n (4, 6, 8...) generations were used to construct the gene pool that is used to determine the intermediate region used for crossover and mutation operator.

For crossover, order-based and partially matched crossover methods were used. In order-based method, a random number of solution points are selected from parent chromosomes. In the first chromosome, the selected gene will remain at its place whereas, in the second chromosome, the corresponding gene will be besides that of the first chromosome that occupies the same place. Order-based crossover method is shown in Figs. 5.4 and 5.5. Figure 5.4 shows the selected chromosomes and Fig. 5.5 shows the resultant chromosomes.

In the Partially Matched Method (PMX), two crossover points are randomly selected to give matching selection. Position wise exchange takes place then. It is also called partially mapped crossover as parents are mapped to each other. Figures 5.6 and 5.7 show the process of partially mapped crossover method.

Fig. 5.4 Selected chromosomes for order-based crossover method

1	3	5	6	7	9	12
2	4	9	8	10	11	6

Fig. 5.5 Chromosomes resulted after order-based crossover operator

1 2 5 6 7 10 12

2 1 9 8 10 7 6

Fig. 5.6 Selected chromosomes for partial mapped crossover method

1 2 3 | 5 4 6 7 | 8 9

4 5 2 | 1 8 7 6 | 9 3

Fig. 5.7 Chromosomes
resulted after crossover
operator

$$8\ 1\ 2\ |\ 5\ 4\ 6\ 7\ |\ 9\ 3$$

$$5\ 2\ 3\ |\ 1\ 8\ 7\ 6\ |\ 4\ 9$$

Fig. 5.8 Inversion mutation
method

$$2\quad 5\quad 6\ |\ 8\quad 10\quad 12\quad 14\ |\ 15\quad 17$$

$$2\quad 5\quad 6\ |\ 14\quad 12\quad 10\quad 8\ |\ 15\quad 17$$

Fig. 5.9 Adjacent
two-change mutation method

$$2\quad 5\quad 6\quad \boxed{8\quad 10}\quad 12\quad 14\quad 15\quad 17$$

$$2\quad 5\quad 6\quad \boxed{10\quad 8}\quad 12\quad 14\quad 15\quad 17$$

For mutation, inversion and two-change mutation operators were used. In inversion method, a subtour is randomly selected by determining two points in chromosome and genes are inverted between selected points whereas in adjacent two input change mutation method, adjacent two genes are selected and the place of genes are inverted. Figures 5.8 and 5.9 show both mutation methods.

5.4   Incremental Feature Selection Algorithm (IFSA)

Qian et al. [5] present an Incremental Feature Selection Algorithm (IFSA) for feature subset selection. It starts with an original feature subset P. It then incrementally computes the new dependency function and evaluates P to find if it is the required feature subset or not. If the new dependency function under P is equal to that under the whole feature set, P is also the new feature subset; otherwise, a new feature subset is computed from P. Algorithm proceeds to gradually select features with the highest significance from C to P and adds them to feature subset. At the final stage, the algorithm removes the redundant features to ensure optimal feature subset output. Finally, redundant features are removed to ensure the optimal output in redundancy removing step. The proposed solution compares feature significance to select the surviving features. The algorithm uses the following definitions to measure the significance of an attribute:

Definition 1 Let DS = (U, A = C $\cup$ D) be a decision system, for B $\subseteq$ C and a $\in$ B. The significance measure of attribute 'a' is defined by $sig_1(a, B, D) = \gamma B$ (D) $- \gamma_{B-\{a\}}(D)$.
If $sig_1(a, B, D) = 0$, then the feature 'a' can be removed otherwise not.

Definition 2 Let DS = (U, A = C $\cup$ D) be a decision system for B $\subseteq$ C and a $\notin$ B. The significance of feature 'a' is defined by $sig_2(a, B, D) = \gamma B \cup \{a\}(D) - \gamma B(D)$.
Figure 5.10 shows the pseudocode of the algorithm.

Input A decision system DS = (U, A = C U D), the original feature subset Red, the original position region $\gamma c(D)$, and the adding feature set C_{ad} or the deleting feature set C_{de}, where $C_{ad} \cap C = \emptyset$, $C_{ad} \subseteq C$;
Output A new feature subset Red'.

Begin
   1) Initialize P ← Red;
   2) If a feature set C_{ad} is added into the system DS;
   3) Let C' ← C U C_{ad};
   4) Compute the equivalence classes U/ C' and $\gamma c'$ (D);
     //According to Theorem 1
   5) for i = 1 to | C_{ad} | do
   6) compute $sig1(c_i, C_{ad}, D)$;
   7) if $sig1(c_i, C_{ad}, D) > 0$, then P ← P U $\{c_i\}$;
   8) end for
   9) if $\gamma p(D) = \gamma c'(D)$, turn to Step 25; else turn to Step 16;
  10) End if
  11) If a feature set C_{de} are deleted from the system DS;
  12) Let C' ← C - C_{de};
  13) if $C_{de} \cap P = \emptyset$, turn to step 25; else P ← P - C_{de} and turn to step 14;
  14) Compute the equivalence classes U/ C' and $\gamma c'$ (D);
     //According to Theorem 2
  15) End if
  16) For $\forall c \in$ C' - P, construct a descending sequence by sig2(c, P, D), and record the result
     by $\{ c'_1, c'_2, ..., c'_{|C'-P|} \}$;
  17) While $(\gamma p(D) \neq \gamma c'(D))$ do
  18) for j = 1 to | C' - P | do
  19) select P ← P U $\{ c'_j\}$ and compute $\gamma p(D)$;
  20) End while
  21) For each $c_j \in$ P do
  22) compute $sig1(c_j, P, D)$;
  23) if $sig1(c_j, P, D) = 0$, then P ← P - $\{c_j\}$;
  24) end for
  25) Red' ← P, return Red';
End

Fig. 5.10 IFSA taken from [5]

5.5 Feature Selection Method Using Fish Swarm Algorithm (FSA)

Chen et al. [6] present a rough set based feature selection method using Fish Swarm Algorithm (FSA). In the first step, the algorithm constructs the initial swarm of fish with each fish, searches for food, and represents a subset of features. With the passage of time, these fish change their position to search for food, communicate with each other to find a locally and globally best position, and the position with a minimum high density of the food. After a fish achieves maximum fitness, it perishes by getting rough set Reduct. The next iteration starts after all the fish perish. The process continues until it gets the same Reducts in three consecutive iterations or maximum iteration threshold is met. Figure 5.11 shows the flow of the FSA process.

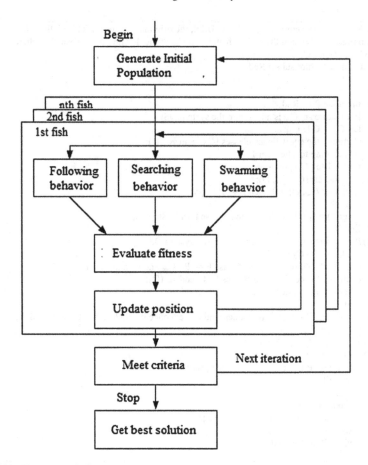

Fig. 5.11 Flowchart of FSA

Some underlying concepts that must be considered before applying FSA to feature selection are the following.

5.5.1 Representation of Position

A fish's position is represented by a binary bit string of length N, where N is the total number of features. The presence of a feature in a fish is represented by binary bit '1' and absence of a feature is represented by '0'. For example, if N = 5 the following fish shown in Fig. 5.12 represents the presence of first, third, and fourth feature from the dataset.

Fig. 5.12 A sample fish

1	0	1	1	0

5.5.2   Distance and Center of Fish

Suppose two fish are represented by two-bit strings X, Y representing the position of these two fish, the hamming distance will be calculated by X XOR Y, i.e., the number of bits at which strings are different. Mathematically,

$$h(X, Y) = \sum_{i=1}^{N} x_i \oplus y_i,$$

where '$\oplus$' is modulo-2 addition, x_i; y_i {0, 1}. The variable x_i represents a binary bit in X.

5.5.3   Position Update Strategies

In each iteration, every fish starts with a random position. Fish change their position one step according to searching, swarming, and following behavior. The authors have used fitness function to evaluate all of these behaviors. The behavior with maximum fitness value updates the next position.

5.5.4   Fitness Function

The following fitness function was used in the algorithm:

$$\text{Fitness} = \alpha * \gamma_R(D) + \beta * \frac{|C| - |R|}{|C|},$$

where $\gamma_R(D)$ is the dependency of decision attribute 'D' on 'R' and R is the number of '1' bits in a fish, and |C| is the number of features in the dataset.

5.5.5   Halting Condition

When a fish achieves maximum fitness, it is perished by getting a rough set Reduct. Next iteration starts after all the fish perish. The algorithm stops when the maximum iteration threshold is met or the same feature Reduct is obtained under three consecutive iterations.

5.6   Feature Selection Method Based on Quick Reduct and Improved Harmony Search Algorithm (RS-IHS-QR)

Inbarani et al. [7] propose a feature selection method based on QuickReduct and Improved Harmony Search Algorithm (RS-IHS-QR). This algorithm emulates the music improvisation process, where each musician improvises their instrument's pitch by searching for a perfect state of harmony. The algorithm stops when it reaches the maximum number of iterations or finds a harmony vector with maximum fitness. It uses a rough set based dependency measure as its objective function to measure the fitness of harmony vector, which again is a performance bottleneck for larger datasets.

5.7   A Hybrid Feature Selection Approach Based on Heuristic and Exhaustive Algorithms Using Rough Set Theory (FSHEA)

In [8], we have proposed a new solution for feature selection based on relative dependency. Their approach is a two-step process:

Step 1: A preprocessor that finds an initial feature subset. For this step, we have used a genetic algorithm and PSO for the sake of experiments. Any heuristic algorithm, e.g., FSA, ACO, etc., can be used.

Step 2: Feature subset optimization. This step optimizes the initial solution produced by Step-1, by removing unnecessary features. Here, any of the exhaustive feature subset algorithms can be used. We have used the one based on relative dependency, as it avoids the calculation of computationally expensive positive region.

Figure 5.13 shows the diagrammatic representation of the proposed solution.

The proposed approach has changed the role of the algorithms used in any of the abovementioned steps. Instead of being used to find the actual Reduct set, heuristic algorithms find an initial Reduct set, which is then refined by using exhaustive search. The following section provides a brief description of each of these methods along with their role in the hybrid approach.

5.7.1   Feature Selection Preprocessor

Preprocessor is the first step of the proposed solution, which provides us with initial candidate Reduct set. At this stage, we have used heuristic algorithms. The reason behind is that the heuristic algorithms help us find the initial feature subset within

Fig. 5.13 Proposed hybrid
approach for feature selection

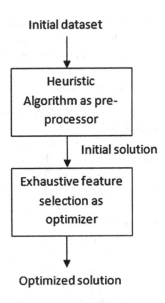

minimal time without exhaustively searching entire dataset. However, the heuristic algorithms do not ensure optimal solution, so the feature subset produced by pre-processor may contain unnecessary features as well, which are then removed in the optimization step. At this stage, any type of heuristic algorithms can be used. For the sake of this paper, we have used a genetic algorithm and particle swarm optimization approach. Below is the brief description of both of these algorithms along with slight changes made for the purpose of enhancing computational efficiency.

5.7.1.1   Genetic Algorithm (GA) as Preprocessor

Genetic algorithm is one of the heuristic-based algorithms, where candidate solutions are deployed in the form of chromosomes. Each of the chromosomes comprises of the genes. In case of feature subset selection, a gene may represent the presence of an attribute. Complete detail of GA can be found in [8, 9]. Here is a description of the points where we made slight changes in this algorithm.

Chromosome encoding: In the conventional genetic algorithm, the chromosome comprises the set of genes, each of which is randomly selected and represents a part of the solution. We, however, have avoided the random gene selection. Instead, chromosomes were encoded in such a way all the attributes of the datasets are present, e.g., if the dataset has 25 attributes, then a particular generation may have five chromosomes each having five genes and each gene representing an attribute. This step was taken for the following two reasons:

(a) Selecting the chromosome size is a problem in GA. To let GA be independent of the size, we encoded all the attributes as genes so that if any size of the chromosome is chosen, all attributes are tested.

(b) As genes are randomly selected, so there is a chance that some attributes are skipped. Encoding all the attributes ensures that all the attributes are tested.

Note that any other encoding scheme can be used in the proposed hybrid approach.

Crossover: For the purpose of our algorithm, we have used a conventional one-point crossover. However, the crossover was performed in decreasing order of relative dependency. This was based on our observation that a chromosome C_1 with a higher order of relative dependency, if crossovers with another higher order relative dependency chromosome, it is more likely that the produced offsprings will have high relative dependency value as compared to the one, if C_1 crosses over with a lower relative dependency value chromosome. Using the abovementioned crossover, order is more likely to produce higher relative dependency offsprings, thus resulting in a fewer number of generations.

Fitness function: The fitness function comprised of the relative dependency of the attributes represented by genes present in the current chromosome. Relative dependency was calculated using equation mentioned in [8]. The first chromosome with the highest fitness value, i.e. one (1) was chosen as a candidate solution.

5.7.1.2   Particle Swarm Optimization (PSO) Algorithm as Preprocessor

We also used particle swarm algorithm as a preprocessor. It is another heuristic-based approach based on swarm logic. A particle in a swarm represents a potential candidate solution. Each swarm has a local best represented by pbest.

The algorithm evaluates the fitness of all particles and the one having the best fitness becomes local best. After finding local best, PSO attempts to find global best particle, i.e., gbest. Gbest particle is the one having the best fitness so far throughout the swarm. PSO then attempts to update the position and velocity of each particle based on its position and distance from local best. The algorithm can be deployed for feature selection using rough set based conventional dependency measure, as implemented in [11], in which case fitness of each particle was measured by calculating dependency of decision attribute on the set of conditional attributes, represented by this particle.

However, we have slightly modified the fitness function and have replaced a rough set based dependency measure with relative attribute dependency. The reason behind is that the relative dependency measure intends to avoid the computationally expensive positive region, thus increasing the computational efficiency of the algorithm. Any other feature selection algorithm can be used here.

5.7.2   Using Relative Dependency Algorithm to Optimize the Selected Features

Preprocessor step provides an initial Reduct set. However, as inherent with heuristic-based algorithms, the selected attributes may have many irrelevant attributes. We have used the relative dependency algorithm to optimize the generated Reduct set. Applying the relative dependency algorithm at this stage, however, does not degrade the performance as the input set for the algorithm has already been reduced by the preprocessor. The algorithm evaluates the Reduct set in a conventional way to find out, if there is any irrelevant feature.

Using the hybrid approach provides us with the positive features of both approaches. The following are some advantages of the proposed hybrid approach:

- Using heuristic algorithm lets us produce the candidate Reduct set without going deep in exhaustive search, which results in cutting down of the execution time, and thus makes use of the proposed approach possible for average and large data sets.
- Using relative dependency as optimizer lets us avoid the calculation of computationally expensive positive region, which prohibits the use of conventional rough set based dependency measure for feature selection in datasets beyond smaller size.
- Using relative dependency algorithm, after preprocessor, lets us further optimize the generated Reduct set. So, it will be the set of minimum possible attributes that can be considered as the final required Reduct set. An important point to note here is that the relative dependency algorithm, at this stage, does not affect the execution time because the input dataset has already been reduced by Step-1.

The proposed solution provides a unique and novel way to perform feature selection. It differs from tradition feature selection algorithms by extracting the positive features of both exhaustive and heuristic-based algorithms instead of using them standalone. Both of these search categories discussed in Sect. 5.1 have strong and weak points, e.g., exhaustive search is best to find the optimal solution as compared to other categories, however, it requires a lot of resources and in case of larger problem space, it becomes totally unpractical. So, for feature selection in larger datasets, exhaustive search is not a good option. The proposed solution, unlike the other algorithms, uses its strength to optimize the final result instead of using it directly on the entire dataset. So instead of using the entire set of attributes as input for exhaustive search algorithm, it is given the reduced set which is already a candidate solution. So, unlike other algorithms, the role of exhaustive search is reduced to optimize the already found solution instead of finding it.

On the other hand, heuristic-based search does not dig deep into problem space to try every solution. The ultimate consequence is that it does not ensure the optimal result set, e.g., in case of using GA for feature selection, it can find the best chromosome in first n iterations, however, does this chromosome represent the minimal feature set? You cannot ensure. However, this feature enables heuristic-based approaches to be used for larger datasets. The proposed, in contrast to other feature selection approaches, uses heuristic-based algorithms to find the initial feature set rather than the final solution. So, the role of heuristic-based algorithm is limited to find the initial solution instead of the final one, which enables the proposed solution best option for larger datasets.

With the above facts, it can be said that the proposed hybrid approach is suitable for feature selection in large data sets to find an optimal solution. Experimental results have justified our proposed approach, both in terms of efficiency and effectiveness.

In [4], the authors have proposed a novel feature selection approach using rough set theory based on hit and trial search method (FSRFV). The algorithm comprises two steps:

1. Construct a feature vector from input dataset using a random selection of attributes using rough set based dependency measure.
2. Optimize the feature vector from the first step by removing the irrelevant features to output the optimal feature subset.

Here, we provide the details of each step.

Step 1: In the first step, the algorithm constructs a feature subset by randomly selecting features from the input feature space. Each feature in the dataset has a 50% probability to be selected as part of the output feature subset. The reason behind is to give each feature an equal opportunity. The size of feature vector SFV in this way is equal to the total number of features in the dataset, however, features not present in current feature vector are represented by '0'. Thus, the feature vector '0, 2, 0, 4, 0, 6' shows that feature number '2, 4 and 6' will be included, whereas, feature numbers 1, 3, and 5 are not part of this feature vector.

The algorithm then finds the dependency of the feature vector using rough set based dependency measure. If dependency of the feature vector is equal to that of the entire set of conditional attributes, it means that this feature vector contains the output feature subset, otherwise a new feature subset is calculated using the same mechanism.

The algorithm makes 'n' number of attempts to construct a feature vector, where 'n' can be a user-specified number which can have any suitable value. However, after 'n' number of attempts, if the feature vector is not formed, then the feature vector with maximum dependency is selected.

Step 2: The feature vector constructed in the first step may contain a number of attributes, out of which normally, a few qualify for output feature subset, and the rest are unnecessary. In this step, the feature vector is optimized by removing unnecessary features. A feature is irrelevant and dispensable if removing it does not affect the dependency of the remaining feature vector [8].

The entire feature vector is scanned for all the irrelevant features to ensure the optimal feature subset is generated at the output.

Figure 5.14 shows the pseudocode of the proposed solution.

Figure 5.15 shows the diagrammatic representation of the proposed solution.

Fig. 5.14 Proposed
algorithm

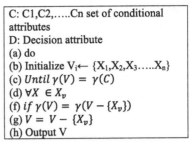

C: C1,C2,.....Cn set of conditional
attributes
D: Decision attribute
(a) do
(b) Initialize $V_i \leftarrow \{X_1,X_2,X_3.....X_n\}$
(c) *Until* $\gamma(V) = \gamma(C)$
(d) $\forall X \in X_v$
(f) *if* $\gamma(V) = \gamma(V - \{X_v\})$
(g) $V = V - \{X_v\}$
(h) Output V

Fig. 5.15 Flowchart of RFS
algorithm

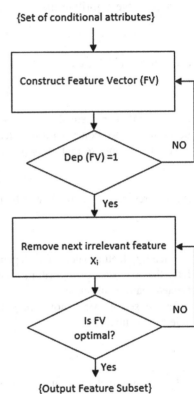

{Set of conditional attributes}

Construct Feature Vector (FV)

Dep (FV) =1   NO

Yes

Remove next irrelevant feature
X_i NO

Is FV
optimal?   NO

Yes

{Output Feature Subset}

We now explain the proposed solution with the help of a simple example. We will consider the dataset given in Table 5.2.

The given decision comprises six objects and five conditional attributes C = {a, b, c, d, e}, where 'a' is first attribute, 'b' is second, and so on. Decision attribute 'Z' comprises two decision classes, i.e., '1' and '2'.

From the given dataset, we know that

$$\gamma(C) = 1.0$$

Table 5.2 Sample decision system

U	A	b	c	d	e	Z
1	L	J	F	X	D	2
2	M	K	G	Y	E	2
3	N	J	F	X	E	1
4	M	K	H	Y	D	1
5	L	J	G	Y	D	2
6	L	K	G	X	E	2

Suppose the algorithm first constructs a first feature vector:

$$V = \{a, b, 0, d, 0\}$$

i.e., features 'a', 'b', and 'd' will be included. Now

$$\gamma(V) = 0.4$$

Here, $\gamma(V)$ is not equal to $\gamma(C)$, so, this feature vector does not contain feature subset, hence, we construct a new feature vector.

This time suppose that algorithm constructs the following feature vector:

$$V = \{a, 0, c, d, 0\}$$

i.e., features 'a', 'c', and 'd' will be included. Now

$$\gamma(V) = 1.0$$

Since this feature vector has dependency '1.0', so it contains the feature subset. Here, the first step of algorithm completes and the feature vector {a,c,d} is the selected feature subset.

As the second step, now we will optimize this feature vector by removing irrelevant features.

$$\gamma(V) - \gamma(V - \{d\}) = 1.0 - 1.0 = 0.$$

This means that removing the feature 'd' from feature vector does not affect dependency, so it is irrelevant and can be removed. Hence, feature vector contains attributes: {a, c}. Now,

$$\gamma(V) - \gamma(V - \{c\}) = 1.0 - 0.4 = 0.6$$

i.e., removing the feature 'c' reduces dependency, so, feature 'c' is indispensable and cannot be removed. Next, we consider the removal of feature 'a'.

$$\gamma(V) - \gamma(V - \{a\}) = 1.0 - 0.4 = 0.6$$

Again, feature 'a' is also indispensable and cannot be removed. So, our final feature subset will comprise of the features:

$$V = \{a, c\}.$$

The proposed feature selection method presents a novel feature selection approach without the use of any complex operators. The following are a few advantages of the proposed solution:

1. There are many approaches in the literature that are based on random feature selection, e.g., genetic algorithms, particle swarm based approaches, fish swarm based approaches, etc. Selecting features in this way lets them avoid exhaustive search, but one of the major drawbacks is that they do not provide an optimal solution, as the resulting output subset may contain many irrelevant features that are also selected. The proposed approach removes the irrelevant feature, thus ensuring the optimal feature subset as output.
2. The proposed solution does not make use of any complex operators, e.g., mutation, crossover, in genetic algorithms, local best and global best in particle swarm and fish swarm algorithms, etc., which results in enhanced performance of the algorithm as compared to these approaches.

5.8 A Rough Set Based Feature Selection Approach Using Random Feature Vectors

In [10], the authors have proposed a novel feature selection approach using rough set theory based on random feature vector selection method (FSRFV). The algorithm comprises two steps:

1. Construct a feature vector from input dataset using a random selection of attributes using rough set based dependency measure.
2. Optimize the feature vector from the first step by removing the irrelevant features to output the optimal feature subset.

Here is a detail of each step.

Step 1: In the first step, the algorithm constructs a feature subset by randomly selecting features from the input feature space. Each feature in the dataset has a 50% probability to be selected as part of the output feature subset. The reason behind is to give each feature an equal opportunity. The size of feature vector SFV in this way is equal to the total number of features in the dataset, however, features not present in current feature vector are represented by '0'. Thus, the feature vector '0, 2, 0, 4, 0, 6' shows that feature number '2, 4, and 6' will be included whereas feature numbers 1, 3, and 5 are not part of this feature vector.

The algorithm then finds the dependency of the feature vector using the conventional dependency measure. If dependency of the feature vector is equal to that of the entire set of conditional attributes, it means that this feature vector contains

Fig. 5.16 Proposed FSRFV
algorithm

C: C1,C2,.....Cn set of conditional
attributes
D: Decision attribute
(a) do
(b) Initialize $V_i \leftarrow \{X_1,X_2,X_3.....X_n\}$
(c) *Until* $\gamma(V) = \gamma(C)$
(d) $\forall X \in X_v$
(f) *if* $\gamma(V) = \gamma(V - \{X_v\})$
(g) $V = V - \{X_v\}$
(h) Output V

the output feature subset, otherwise, we calculate a new feature subset using the same mechanism.

The algorithm makes 'n' number of attempts to construct a feature vector, where 'n' can be a user-specified number which can have any suitable value. However, after 'n' number of attempts, if the feature vector is not formed, then the feature vector with maximum dependency is selected.

Step 2: The feature vector constructed in the first step may contain a number of attributes, out of which normally a few qualify for output feature subset, rest are unnecessary. In this step, the feature vector is optimized by removing unnecessary features. A feature is irrelevant and dispensable if removing it does not affect the dependency of the remaining feature vector [26].

The entire feature vector is scanned for all the irrelevant features to ensure the optimal feature the subset is generated at the output. Figure 5.16 shows the pseudodocode of the proposed solution.

We now explain the proposed solution with the help of a simple example. We will consider the dataset given in Table 5.2.

The given decision comprises six objects and five conditional attributes C = {a, b, c, d, e}, where 'a' is first attribute, 'b' is second, and so on. Decision attribute 'Z' comprises two decision classes which are '1' and '2'.

From the given dataset, we know that

$$\gamma(C) = 1.0$$

Suppose the algorithm first constructs a feature vector:

$$V = \{a, b, 0, d, 0\}$$

So, features 'a', 'b', and 'd' will be included. Now

$$\gamma(V) = 0.4$$

Here, $\gamma(V)$ is not equal to $\gamma(C)$, so, this feature vector does not contain feature subset, hence we construct a new feature vector.

This time suppose that algorithm constructs the following feature vector:

$$V = \{a, 0, c, d, 0\}$$

So, features 'a', 'c', and 'd' will be included. Now

$$\gamma(V) = 1.0$$

As this feature vector has dependency '1.0', so it contains the feature subset. Here, the first step of algorithm completes and the feature vector {a, c, d} is the selected feature subset.

As the second step, now we will optimize this feature vector by removing irrelevant features.

$$\gamma(V) - \gamma(V - \{d\}) = 1.0 - 1.0 = 0$$

This means that removing the feature 'd' from feature vector does not affect dependency, so it is irrelevant and can be removed. Hence, feature vector contains attributes: {a, c}. Now,

$$\gamma(V) - \gamma(V - \{c\}) = 1.0 - 0.4 = 0.6$$

Here, removing the feature 'c' reduces dependency, so, feature 'c' is indispensable and cannot be removed. Next, we consider the removal of the feature 'a'.

$$\gamma(V) - \gamma(V - \{a\}) = 1.0 - 0.4 = 0.6$$

Again, the feature 'a' is also indispensable and cannot be removed. So, our final feature subset will comprise the features:

$$V = \{a, c\}$$

The proposed feature selection method presents a novel feature selection approach without the use of any complex operators. The following are a few advantages of the proposed solution:

1. There are many approaches in the literature that are based on random feature selection, genetic algorithms, particle swarm based approaches, fish swarm based approaches are a few to name. Selecting features in this way lets them avoid exhaustive search, but one of the major drawbacks is that they do not provide an optimal solution, as the resulting output subset may contain many irrelevant features that are also selected. The proposed approach removes the irrelevant feature, thus ensuring the optimal feature subset as output.

2. The proposed solution does not make use of any complex operators like mutation and crossover in genetic algorithms, local best and global best in particle swarm and fish swarm algorithms. It results in enhanced performance of the algorithm as compared to these approaches.

The authors claim to produce optimized results as compared to other feature vectors algorithms.

5.9 Heuristic-Based Dependency Calculation Technique

In [11], Raza et al., have presented a heuristic-based approach for feature selection using Rough Set Theory. The approach was used to find consistent records regarding each decision class in the dataset. This technique allows calculating dependency by avoiding the positive region, which ultimately enhances the computational efficiency of the underlying feature selection algorithm thus enabling it to be used for the dataset

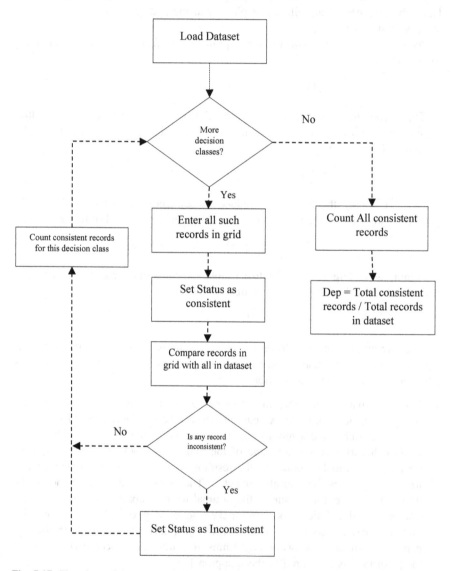

Fig. 5.17 Flowchart of the proposed HDC-based dependency calculation method

beyond smaller size. In order to calculate dependency by using the proposed method, the authors have used a two-dimensional grid as an intermediate data structure.

Figure 5.17 shows the flowchart of the proposed approach.

5.10   Parallel Dependency Calculation Method for Feature Selection

In [12], Raza et al., have presented a parallel dependency calculation technique for feature selection. In Rough Set Theory, dependency measure is the main criteria for selection of features, so, accomplishing the calculation of dependency using parallel approach may enhance the efficiency of the underlying algorithm using this measure.

Figures 5.18 and 5.19 show the flowchart of the proposed approach.

Table 5.3 shows the summary of all the rough set based approaches discussed so far.

5.11   Summary

In this section, we have presented feature selection algorithms using rough set based positive region and alternate ones. Positive-region-based approaches use conventional dependency measure comprising three steps to measure the fitness of an attribute for being selected for Reduct set. However, using positive region is a

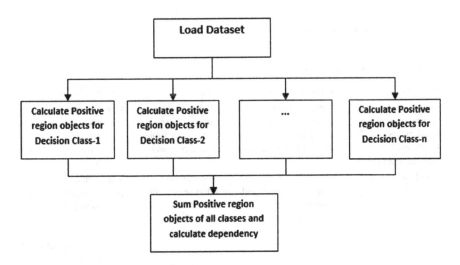

Fig. 5.18 Flowchart of PDC-based technique

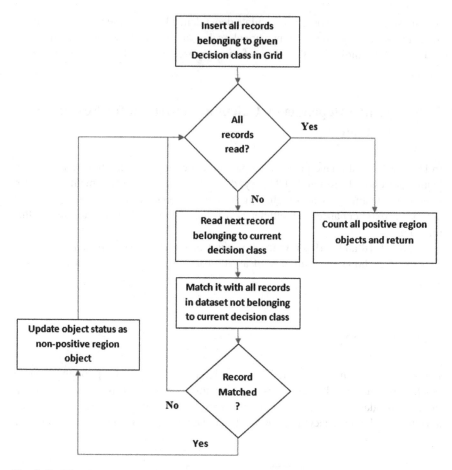

Fig. 5.19 Flowchart of how each process works

Table 5.3 Related algorithms based on RST

Algorithm	Technique used	Advantages	Disadvantages
Supervised hybrid feature selection based on PSO and rough sets for medical diagnosis [3]	Particle swarm optimization and rough set based dependency measure	PSO is an advanced heuristic-based algorithm to avoid exhaustive search	Conventional dependency-based measure is a performance bottleneck
Rough set based genetic algorithm [4]	Conventional positive-region-based dependency calculation	It is based on randomness so the procedure may find Reducts in few attempts	Uses conventional positive-region-based dependency measure

(continued)

Table 5.3 (continued)

Algorithm	Technique used	Advantages	Disadvantages
Quick Reduct approach for feature selection [1]	Rough set based dependency measure	Attempts to calculate Reducts without exhaustively generating all possible subsets	Uses conventional dependency-based measure, which is a time-consuming task
REVERSEREDUCT [1]	Rough set based dependency measure	Backward elimination is utilized without exhaustively generating all possible combinations	Dependency is calculated using conventional positive-region-based approach
An incremental algorithm to feature selection in decision systems with the variation of feature set [5]	Incremental feature selection using rough set based significance measure of attributes	Presents feature selection for dynamic systems where the datasets keep on increasing with time	Conventional dependency measure is used to measure attribute significance. Measuring significance requires measuring dependency twice, once with attribute and then without the attribute
Fish swarm algorithm [6]	Rough set based dependency used with fish swarm method for feature selection	Attempts to find rough set Reducts using the swarm logic where swarms can discover the best combination of features	Conventional dependency-based measure is again a performance bottleneck
Rough set improved harmony search QuickReduct [7]	Rough set based dependency measure used with harmony search algorithm for feature selection	Integrates rough set theory with 'improved harmony search' based algorithm with QuickReduct for feature selection	Uses conventional dependency-based measure, which is time consuming
Feature selection using rough set based heuristic dependency calculation [11]	Heuristics dependency calculation	Improves performance by avoiding the positive region	Can be used for supervised feature selection only
A parallel rough set based dependency calculation method for efficient feature selection [12]	Parallel dependency calculation technique	Improves performance by decreasing the execution time	Can only be used for supervised feature selection

computationally expensive approach that makes these approaches inappropriate to use for larger datasets. Alternate approaches do not use positive region. However, the application of such approaches has only been tested against smaller datasets, which raise the question for their appropriateness for larger datasets.

References

1. Jensen R, Shen Q (2008) Computational intelligence and feature selection: rough and fuzzy approaches, vol 8. Wiley
2. Pethalakshmi A, Thangavel K (2007) Performance analysis of accelerated QuickReduct algorithm. In: International Conference on Computational Intelligence and Multimedia Applications, 2007, vol 2. IEEE
3. Inbarani HH, Azar AT, Jothi G (2014) Supervised hybrid feature selection based on PSO and rough sets for medical diagnosis. Comput Methods Programs Biomed113(1):175–185
4. Zuhtuogullari Kursat, Allahverdi Novruj, Arikan Nihat (2013) Genetic algorithm and rough sets based hybrid approach for reduction of the input attributes in medical systems. Int J Innov Comput Inf Control 9:3015–3037
5. Qian W, et al (2015) An incremental algorithm to feature selection in decision systems with the variation of feature set. Chin J Electron 24(1):128–133
6. Chen Yumin, Zhu Qingxin, Huarong Xu (2015) Finding rough set reducts with fish swarm algorithm. Knowl-Based Syst 81:22–29
7. Inbarani HH, Bagyamathi M, Azar AT (2015) A novel hybrid feature selection method based on rough set and improved harmony search. Neural Comput Appl 26(8):1859–1880
8. Raza MS, Qamar U (2016) A hybrid feature selection approach based on heuristic and exhaustive algorithms using Rough set theory. In: Proceedings of the International Conference on Internet of things and Cloud Computing. ACM
9. Raza MS, Qamar U (2017) Feature selection using rough set based heuristic dependency calculation. PhD dissertation, NUST
10. Raza MS, Qamar U (2016) A rough set based feature selection approach using random feature vectors. In: 2016 International Conference on Frontiers of Information Technology (FIT). IEEE
11. Raza MS, Qamar U (2018) A heuristic based dependency calculation technique for rough set theory. Pattern Recognit 81:309–325
12. Raza MS, Qamar U (2018) A parallel rough set based dependency calculation method for efficient feature selection. Appl Soft Comput 71:1020–1034

Chapter 6
Unsupervised Feature Selection Using RST

Supervised feature selection evaluates the features that provide maximum of the information based on classification accuracy. This requires labeled data, however, in real world not all the data is properly labeled, so we may come across the situation where little or no class information is provided. For such type of data, we need unsupervised feature selection information that could find feature subsets without given any class labels. In this section, we will discuss some of the unsupervised feature subset algorithms based on Rough Set Theory.

6.1 Unsupervised Quick Reduct Algorithm (USQR)

In [1], the authors have presented an unsupervised QuickReduct algorithm using RST. Original QuickReduct algorithm, which is supervised, takes two inputs, i.e. set of conditional attributes and decision attribute(s). However, USQR takes only one input, i.e. set of conditional attributes. However, just like the existing QuickReduct algorithm, it performs feature selection without exhaustively generating all possible subsets.

It starts with an empty set and adds those attributes one by one which results in the maximum degree of increase in dependency until it produces maximum possible value. According to the algorithm, the mean dependency of each attribute subset is calculated and the best candidate is chosen:

$$\gamma_p(a) = \frac{|POS_p(a)|}{|U|}, \forall a \in A.$$

Following is the pseudocode of the algorithm (Fig. 6.1).

Now we explain USQR with an example taken from [1]. We will consider the following dataset (Table 6.1).

© Springer Nature Singapore Pte Ltd. 2019
M. S. Raza and U. Qamar, *Understanding and Using Rough Set Based Feature Selection: Concepts, Techniques and Applications*, https://doi.org/10.1007/978-981-32-9166-9_6

Fig. 6.1 Unsupervised
QuickReduct algorithm

USQR(C)
C, the set of all conditional features;
(1) R ← {}
(2) do
(3) T← R
(4) ∀ x ∈ (C − R)
(5) ∀ y ∈ C

(6) $\gamma_{RU(x)}(y) = \frac{|POS_{RU(x)}(y)|}{|U|}$

(7) if $\overline{\gamma_{RU(x)}(y)}, \forall y \in C > \overline{\gamma_r(y)}, \forall y \in C$
(8) T ← RU{x}
(9) R← T
(10) until $\overline{\gamma_R(y)}, \forall y \in C = \overline{\gamma_c(y)}, \forall y \in C$
return R

Table 6.1 Sample dataset
taken from [1]

$x \in U$	a	b	c	d
1	1	0	2	1
2	1	0	2	0
3	1	2	0	0
4	1	2	2	1
5	2	1	0	0
6	2	1	1	0
7	2	1	2	1

Dataset comprises of four conditional attributes, i.e. a, b, c and d. In Step 1, we
calculate the dependency value of each attribute.

Step-1:

$$\gamma_{\{a\}}(\{a\}) = \frac{|POS_{\{a\}}(\{a\})|}{|U|} = \frac{|\{1,2,3,4,5,6,7\}|}{|\{1,2,3,4,5,6,7\}|} = \frac{7}{7}$$

$$\gamma_{\{b\}}(\{b\}) = \frac{|POS_{\{a\}}(\{b\})|}{|U|} = \frac{|\{5,6,7\}|}{|\{1,2,3,4,5,6,7\}|} = \frac{3}{7}$$

$$\gamma_{\{c\}}(\{c\}) = \frac{|POS_{\{a\}}(\{c\})|}{|U|} = \frac{|\{\}|}{|\{1,2,3,4,5,6,7\}|} = \frac{0}{7}$$

$$\gamma_{\{d\}}(\{d\}) = \frac{|POS_{\{a\}}(\{d\})|}{|U|} = \frac{|\{\}|}{|\{1,2,3,4,5,6,7\}|} = \frac{0}{7}$$

$$\sum_{\forall y \in c} \gamma_{\{a\}}(\{y\}) = \frac{7}{7} + \frac{3}{7} + \frac{0}{7} + \frac{0}{7} = \frac{10}{7}$$

$$\overline{\gamma_{\{a\}}(\{y\})}, \forall y \in C = \frac{\frac{10}{7}}{4} = 0.35714$$

Similarly other degrees of dependency values are calculated. Table 6.2 shows these values.

Attribute b generates the highest degree of dependency hence chosen to evaluate the indiscernibility of sets {a, b}, {b, c} and {b, d} and calculate the degree of dependency as given in Step 2.

Step-2:

$$\gamma_{\{a,b\}}(\{a\}) = \frac{|POS_{\{a,b\}}(\{a\})|}{|U|} = \frac{|\{1,2,3,4,5,6,7\}|}{|\{1,2,3,4,5,6,7\}|} = \frac{7}{7}$$

$$\gamma_{\{a,b\}}(\{b\}) = \frac{|POS_{\{a,b\}}(\{b\})|}{|U|} = \frac{|\{1,2,3,4,5,6,7\}|}{|\{1,2,3,4,5,6,7\}|} = \frac{7}{7}$$

$$\gamma_{\{a,b\}}(\{c\}) = \frac{|POS_{\{a,b\}}(\{c\})|}{|U|} = \frac{|\{1,2\}|}{|\{1,2,3,4,5,6,7\}|} = \frac{2}{7}$$

$$\gamma_{\{a,b\}}(\{d\}) = \frac{|POS_{\{a,b\}}(\{d\})|}{|U|} = \frac{|\{\}|}{|\{1,2,3,4,5,6,7\}|} = \frac{0}{7}$$

$$\sum_{\forall y \in c} \gamma_{\{a,b\}}(\{y\}) = \frac{7}{7} + \frac{7}{7} + \frac{2}{7} + \frac{0}{7} = \frac{16}{7}$$

$$\overline{\gamma_{\{a,b\}}(\{y\})}, \forall y \in C = \frac{\frac{16}{7}}{4} = 0.57143$$

Similarly, the other dependency values are shown in Table 6.3.

Table 6.2 Degree of dependency after Step 1

$y|x$	{a}	{b}	{c}	{d}
a	1.0000	1.0000	0.1429	0.0000
b	0.4286	1.0000	0.1429	0.0000
c	0.0000	0.2857	1.0000	0.4286
d	0.0000	0.0000	0.4286	1.0000
$\overline{\gamma_{\{a\}}(\{y\})}, \forall y \in C$	0.3571	0.3571	0.3571	0.3571

Table 6.3 Degree of
dependency after Step 2

$y\vert x$	$\{a, b\}$	$\{b, c\}$	$\{b, d\}$
a	1.0000	1.0000	1.0000
b	1.0000	1.0000	1.0000
c	0.2857	1.0000	0.7143
d	0.0000	0.7143	1.0000
$\gamma_{\{a\}}(\{y\}), \forall y \in C$	0.57143	0.57143	0.57143

Subsets $\{b, c\}$ and $\{b, d\}$ generate highest degree of dependency but algorithm selects $\{b, c\}$ as it appears first and calculates indiscernibility of $\{a, b, c\}$ and $\{b, c, d\}$ as shown below in Step 3:

Step-3:

$$\gamma_{\{a,b\}}(\{a\}) = \frac{|POS_{\{a,b,c\}}(\{a\})|}{|U|} = \frac{|\{1,2,3,4,5,6,7\}|}{|\{1,2,3,4,5,6,7\}|} = \frac{7}{7}$$

$$\gamma_{\{a,b,c\}}(\{b\}) = \frac{|POS_{\{a,b,c\}}(\{b\})|}{|U|} = \frac{|\{1,2,3,4,5,6,7\}|}{|\{1,2,3,4,5,6,7\}|} = \frac{7}{7}$$

$$\gamma_{\{a,b,c\}}(\{c\}) = \frac{|POS_{\{a,b,c\}}(\{c\})|}{|U|} = \frac{|\{1,2\}|}{|\{1,2,3,4,5,6,7\}|} = \frac{2}{7}$$

$$\gamma_{\{a,b,c\}}(\{d\}) = \frac{|POS_{\{a,b,c\}}(\{d\})|}{|U|} = \frac{|\{3,4,5,6,7\}|}{|\{1,2,3,4,5,6,7\}|} = \frac{5}{7}$$

$$\sum_{\forall y \in c} \gamma_{\{a,b,c\}}(\{y\}) = \frac{7}{7} + \frac{7}{7} + \frac{7}{7} + \frac{5}{7} = \frac{26}{7}$$

$$\overline{\gamma_{\{a,b,c\}}(\{y\})}, \forall y \in C = \frac{\frac{16}{7}}{4} = 0.02857$$

Similarly:

$$\overline{\gamma_{\{b,c,d\}}(\{y\})}, \forall y \in C = 1$$

The other dependency values calculated in Step 3 are shown in Table 6.4.

Since the dependency value of subset $\{b, c, d\}$ is one so algorithm terminates and outputs this subset as Reduct.

Table 6.4 Degree of dependency after Step 4

$y \vert x$	$\{a, b, c\}$	$\{b, c, d\}$
a	1.0000	1.0000
b	1.0000	1.0000
c	1.0000	1.0000
d	0.7143	1.0000
$\overline{\gamma_{\{a\}}(\{y\})}, \forall y \in C$	0.9285	1.0000

6.2   Unsupervised Relative Reduct Algorithm

In [2], the authors have presented a relative-dependency-based algorithm for unsupervised datasets (USRelativeDependency). Existing Relative Dependency algorithm using both conditional and decision attributes for feature subset selection. However, USRelativeDependency performs feature selection on the basis of relative dependency using only the conditional set of attributes.

Figure 6.2 shows the pseudocode of the algorithm.

Initial feature subset comprises of all the features in the dataset. It then evaluates each feature. If relative dependency of the feature is '1', it can safely be removed. The relative dependency for unsupervised data can be calculated as follows:

$$K_R(\{a\}) = \frac{|U/IND(R)|}{|U/IND(R \cup \{a\})|}, \forall a \in A$$

Then show that R is a Reduct if and only if $K_R(\{a\}) = K_C(\{a\})$ and $\forall X \subset R, K_X(\{a\}) \neq K_C(\{a\})$. In this case, the decision attribute used in the supervised feature selection is replaced by the conditional attribute a, which is to be eliminated from the current Reduct set R.

Now we will explain unsupervised relative Reduct with and example taken from [2]. We will consider the following dataset (Table 6.5).

Fig. 6.2 Unsupervised QuickReduct algorithm

```
USRelativeReduct(C)
C, the conditional attributes;
    (1)  R← C
    (2)  ∀ a ∈ C
    (3)  if (K_{R−{a}}({a}) == 1)
    (4)  R ← R − {a}
return R
```

Table 6.5 Sample dataset taken from [2]

$x \in U$	a	b	c	d
1	1	0	2	1
2	1	0	2	0
3	1	2	0	0
4	1	2	2	1
5	2	1	0	0
6	2	1	1	0
7	2	1	2	1

As the algorithm uses backward elimination so initially reduct set comprises of the entire set of conditional attributes, i.e. R = {a, b, c, d}. Now the algorithm considers attribute 'a' for elimination:

$$K_{\{b,c,d\}}(\{a\}) = \frac{\left|\frac{U}{IND(b,c,d)}\right|}{\left|\frac{U}{IND(a,b,c,d)}\right|} = \frac{|\{\{1\}\{2\}\{3\}\{4\}\{5\}\{6\}\{7\}\}|}{|\{\{1\}\{2\}\{3\}\{4\}\{5\}\{6\}\{7\}\}|} = \frac{7}{7}$$

As dependency is equal to '1', so attribute 'a' can be safely removed. The Reduct set thus becomes R = {b, c, d}

Now we consider elimination of 'b':

$$K_{\{c,d\}}(\{b\}) = \frac{\left|\frac{U}{IND(c,d)}\right|}{\left|\frac{U}{IND(b,c,d)}\right|} = \frac{|\{\{1,4,7\}\{2\}\{3,5\}\{6\}\}|}{|\{\{1\}\{2\}\{3\}\{4\}\{5\}\{6\}\{7\}\}|} = \frac{4}{7}$$

As dependency is not equal to '1' so we cannot remove 'b'. Next algorithm considers elimination of attribute 'c':

$$K_{\{b,d\}}(\{c\}) = \frac{\left|\frac{U}{IND(b,d)}\right|}{\left|\frac{U}{IND(b,c,d)}\right|} = \frac{|\{\{1\}\{2\}\{3\}\{4\}\{5,6\}\{7\}\}|}{|\{\{1\}\{2\}\{3\}\{4\}\{5\}\{6\}\{7\}\}|} = \frac{6}{7}$$

Again relative dependency does not evaluate to '1' so we cannot eliminate 'c' as well. Now algorithm considers elimination of attribute 'd':

$$K_{\{b,c\}}(\{d\}) = \frac{\left|\frac{U}{IND(b,c)}\right|}{\left|\frac{U}{IND(b,c,d)}\right|} = \frac{|\{\{1,2\}\{3\}\{4\}\{5\}\{6\}\{7\}\}|}{|\{\{1\}\{2\}\{3\}\{4\}\{5\}\{6\}\{7\}\}|} = \frac{6}{7}$$

Again relative dependency is not equal to '1', so we cannot remove 'd'. Thus, Reduct set comprises of attribute R = {b, c, d}.

6.3   Unsupervised Fuzzy-Rough Feature Selection

In [3], the authors have presented an unsupervised fuzzy-rough feature selection algorithm with different fuzzy-rough feature evaluation criteria. The algorithm starts by considering all the features in the dataset. It then evaluates the measure used as evaluation criteria without this feature. If the measure remains unaffected, then the feature is removed. The process continues until no further features can be removed without affecting the corresponding measure.

Following evaluation measures were used:

Dependency Measure: A set of attribute(s) Q depends on the set of attribute(s) P if P uniquely determines Q. However, the authors argue that fuzzy dependency measure along with its use for supervised fuzzy-rough feature selection can also be used for determining the interdependencies between attributes. This can be achieved by replacing the decision class with the set of feature Q.

Boundary Region Measure: Most of the approaches in the crisp-rough set based feature selection and all the approaches of fuzzy-rough feature selection use a lower approximation for feature selection, however, an upper approximation can also be used to discriminate between objects. For example, two subsets may produce the same lower approximation but one may give a smaller upper approximation, means the lesser uncertainty in the boundary region. Similarly, fuzzy boundary region can be used for feature evaluation.

Discernibility Measure: In the case of conventional RST, the feature selection algorithms can be categorized into two broad classes, those using dependency measure and those using discernibility measure. The fuzzy tolerance relations that represent objects' approximate equality can be used to extend the classical discernibility function. For each combination of features P, a value is obtained indicating how well these attributes maintain the discernibility, relative to another subset of features Q, between all objects.

The pseudocode of the algorithm is given in Fig. 6.3.

The algorithm can be used by specifying any of the evaluation measure mentioned above. The complexity of the search in the worst case is O(n), where n is the number of original features.

Fig. 6.3 Unsupervised QuickReduct algorithm

```
UFRQUICKREDUCT(F)
F, the set of all features.
    (1) R ← C
    (2) foreach x ∈ ℂ
    (3) R ← R − {x}
    (4) if M(R, {x}) < 1
    (5) R ← R U {x}
    (6) return R
```

6.4   Unsupervised PSO Based Relative Reduct (US-PSO-RR)

The authors in [4] have presented a hybrid approach for unsupervised feature selection. The approach uses relative Reduct and particle swarm optimization for this purpose. Figure 6.4 shows the pseudocode of their proposed algorithm. The algorithm takes a set of conditional attributes 'C' as input and produces the Reduct set 'R' as output.

Algorithm: US-PSO-RR(C)
Input: C, the set of all conditional features,
Output: Reduct R

Step 1: Initialize X with random position V_i with random velocity
$\forall: X_i \leftarrow randomPosition();$
$V_i \leftarrow randomVelocity();$
fit $\leftarrow$ 0; globalbest $\leftarrow$ fit;
gbest $\leftarrow X_1$; pbest(1) $\leftarrow X_1$
For i = 1 ... S
pbest(i) = X_i
Fitness(i) = 0
End for

Step 2: While Fitness != 1   // stopping criterion
For i = 1 ... S // for each particle
$\forall: X_i$;//Compute fitness of feature subset of X_i
R $\leftarrow$ Feature subset of X_i (1's of X_i)
$\forall a \in (y)$
$$\gamma_R(a) = \frac{|U/IND(R)|}{|U/IND(RU\{a\})|}$$
$Fit = \overline{\gamma_R}(y) \forall y \notin R$
if Fitness(i) > fit
Fitness(i) = fit
pbest(i) = X_i
End
if(Fit == 1)
return R
End if
End for

Step 3: Compute best fitness
For i = 1, ...,S
if(Fitness(i) > globalbest)
gbest $\leftarrow X_i$;
globalbest $\leftarrow$ Fitness(i);
End if
End for
UpdateVelocity(); // Update velocity V_i's of X_i's
UpdatePosition(); // Update position of X_i's, Continue with the next iteration
End {while}
Output Reduct R

Fig. 6.4 Pseudo code of US-PSO-RR

In Step 1, the proposed algorithm initializes the initial populations of particles with randomly selected conditional attributes and initial velocity. A population of particles is constructed then. For each particle, mean relative dependency is calculated. If the dependency is '1', it is considered as Reduct set. If mean dependency is not equal to '1', then the pbest (highest relative dependency value) of each particle is retained and the best value of the entire population is retained as the global best value. Algorithm finally updates position and velocity the next population is generated.

Encoding:

Algorithm uses '1' and '0' to represent the presence and absence of attributes. The attribute that will be included as part of particle position is represented by '1' and the particle which will not be part of the particle will be represented by '0'. For example, if there are five attributes, i.e. a, b, c, d and e and we need to include the attributes b, c and e, the particle will be like

a	b	c	d	e
0	1	1	0	1

The above particle position shows that particle will include attribute b, c, and e whereas attributes a and d will be absent.

Representation and Updation of Velocity and Positions:

Velocity of particles is represented by positive integer between 1 and V_{max}. It basically determines how many bits of the particle should be changed with respect to the global best position. Pbest represents the local best and gbest represents the global best index. The velocity of each particle is updated according to the following equation:

$$V_{id} = w * V_{id} + c_1 * rand() * (P_{id} - x_{id}) + c_2 * Rand() * (P_{gd} - x_{id})$$

Here w represents inertia weight and c1 and c2 represent acceleration constants. Particle's position is changed on the basis of velocity as follows:

- If $V \leq xg$, randomly change V bits of the particle, which are different from that of gbest.
- If $V > xg$, change all the different bits to be the same as that of gbest and a further $(V - xg)$ bits should be changed randomly.

W can be calculated as follows:

$$w = w_{max} - \frac{w_{max} - w_{min}}{iter_{max}} iter$$

Here w_{max} is the initial value of the weighting coefficient, w_{min} is the final value, $iter_{max}$ is the maximum number of iterations and iter is the current iteration.

Algorithm uses relative dependency measure to measure the fitness of a particle. Relative dependency is calculated as follows:

$$\gamma_R(a) = \frac{|U/IND(R)|}{|U/IND(R \cup \{a\})|} \forall a \notin R$$

where R is the subset selected by the particle and the mean dependency of selected gene subset, on all the genes that are not selected by the particle is used as the fitness value of the particle X_i.

$$Fitness = Fitness(X_i) = \overline{\gamma_R}(Y) \forall y \notin R$$

We will now explain this algorithm with the help of an example taken from [5]. We will use the following dataset as a sample.

Suppose initially following particle generated was (1, 0, 0, 1). This particle contains the features 'a' and 'd' and excludes feature 'b' and 'c'. Hence R = {a, d} and Y = {b, c}.

Now:

$$\gamma_R(b) = \frac{|INDR_R|}{|INDR_{R \cup \{b\}}|} = \frac{|\{1,4\}\{2,3\}\{5,6\}\{7\}|}{|\{1\}\{2\}\{3\}\{4\}\{5\}\{6\}\{7\}|} = \frac{4}{6} = 0.667$$

$$\gamma_R(c) = \frac{|INDR_R|}{|INDR_{R \cup \{c\}}|} = \frac{|\{1,4\}\{2,3\}\{5,6\}\{7\}|}{|\{1\}\{2\}\{3\}\{4\}\{5\}\{6\}\{7\}|} = \frac{4}{6} = 0.667$$

$$\overline{\gamma_R}(a) \forall a \in Y = \frac{0.667 + 0.667}{2} = 0.667$$

Since $\overline{\gamma_R}(a) \neq 1$, so {a, d} will not be considered as Reduct set. Suppose at some stage a particle Xi = {0, 1, 1, 1} is generated. Dependency will be as follows:

R = {b, c, d} and Y = {a}

$$\gamma_R(a) = \frac{|INDR_R|}{|INDR_{R \cup \{a\}}|} = \frac{|\{1\}\{2\}\{3\}\{4\}\{5\}\{6\}\{7\}|}{|\{1\}\{2\}\{3\}\{4\}\{5\}\{6\}\{7\}|} = \frac{7}{7} = 1$$

$$\overline{\gamma_R}(a) \forall a \in Y = 1$$

So R = {b, c, d} will be Reduct set.

6.5   Unsupervised PSO Based Quick Reduct (US-PSO-QR)

US-PSO-QR works uses the same mechanism as discussed in US-PSO-QR. Figure 6.5 shows pseudocode of the algorithm.

Algorithm takes a set of conditional attributes as input and produces Reduct set 'R' as output. It starts with initial population of particles, evaluates fitness of each particle. A feature with the highest fitness is selected and its combinations with other features are evaluated. The local and global best (pbest and gbest) are updated accordingly. Finally, the algorithm updates the velocity and position of each particle. The process continues until we meet the stopping criteria which normally comprises of maximum number of iterations.

Algorithm: US-PSO-QR(C)
Input: C, the set of features,
Output: Reductset R

Step 1: Initialize X with random position V_i with random velocity
$\forall: X_i \leftarrow randomPosition();$
$V_i \leftarrow randomVelocity();$
fit $\leftarrow$ 0; globalbest $\leftarrow$ Fit;
gbest $\leftarrow X_1;$

Step 2: While Fitness $!= \overline{\gamma_C}(y) \; \forall y \in C$ // stopping criterion
For i = 1 ... S // for each particle
$\forall: X_i; \; T \leftarrow \{\}$
//Compute fitness of feature subset of X_i
$R \leftarrow$ Feature subset of X_i (1's of X_i)
$\forall x \in R; \; \forall y \in C$
$$\gamma_{TU(x)}(y) = \frac{|POS_{TU(x)}(y)|}{|U|}$$
$Fitness(i) = \overline{\gamma_{TU(x)}}(y) \; \forall y \in C$
End for
Step 3: Compute best fitness
For i = 1, ...,S
if(fitness(i) > globalbest)
gbest $\leftarrow X_i$; globalbest $\leftarrow$ Fitness(i); pbest(i) $\leftarrow$ bestPos(X_i);
$if \; fitness(i) = \overline{\gamma_C}(y) \; \forall y \in C$
$R \leftarrow$ getReduct(X_i)
End if
End if
End for
UpdateVelocity(); // Update velocity V_i's of X_i's
UpdatePosition(); // Update position of X_i's, Continue with the next iteration
End {while}
Output Reduct R

Fig. 6.5 Pseudo code of US-PSO-RR

Table 6.6 Sample dataset taken from [5]

$x \in U$	a	b	c	d
1	1	0	2	1
2	1	0	2	0
3	1	2	0	0
4	1	2	2	1
5	2	1	0	0
6	2	1	1	0
7	2	1	2	1

Now we will explain this algorithm with the help of an example. We will consider Table 6.6 and the same initial population.

$$\gamma_{T \cup \{ad\}}(a) = \frac{\left|POS_{T \cup \{ad\}}(a)\right|}{|U|} = \frac{|\{1,2,3,4,5,6,7\}|}{|\{1,2,3,4,5,6,7\}|} = \frac{7}{7} = 1$$

$$\gamma_{T \cup \{ad\}}(b) = \frac{\left|POS_{T \cup \{ad\}}(b)\right|}{|U|} = \frac{|\{5,6,7\}|}{|\{1,2,3,4,5,6,7\}|} = \frac{3}{7} = 0.4286$$

$$\gamma_{T \cup \{ad\}}(c) = \frac{\left|POS_{T \cup \{ad\}}(c)\right|}{|U|} = \frac{|\{1,4,7\}|}{|\{1,2,3,4,5,6,7\}|} = \frac{3}{7} = 0.4286$$

$$\gamma_{T \cup \{ad\}}(d) = \frac{\left|POS_{T \cup \{ad\}}(d)\right|}{|U|} = \frac{|\{1,2,3,4,5,6,7\}|}{|\{1,2,3,4,5,6,7\}|} = \frac{7}{7} = 1$$

$$\overline{\gamma_{T \cup \{ad\}}}(y); \forall y \in C = \frac{1 + 0.4286 + 0.4286 + 1}{4} = 0.7143$$

$$\overline{\gamma_{T \cup \{bd\}}}(y); \forall y \in C = \frac{\left|POS_{T \cup \{bd\}}(d)\right|}{|U|} = 0.9286$$

$$\overline{\gamma_{T \cup \{ab\}}}(y); \forall y \in C = \frac{\left|POS_{T \cup \{ab\}}(d)\right|}{|U|} = 0.5714$$

$$\overline{\gamma_{T \cup \{bc\}}}(y); \forall y \in C = \frac{\left|POS_{T \cup \{bc\}}(d)\right|}{|U|} = 0.9286$$

Since none of the attributes shows dependency of '1', so the next iteration starts. Suppose at any stage the particle is (0, 1, 1, 1), dependency will be calculated as

$$\gamma_{T \cup \{bcd\}}(a) = \frac{\left|POS_{T \cup \{bcd\}}(a)\right|}{|U|} = \frac{|\{1,2,3,4,5,6,7\}|}{|\{1,2,3,4,5,6,7\}|} = \frac{7}{7} = 1$$

$$\gamma_{T \cup \{bcd\}}(b) = \frac{|POS_{T \cup \{bcd\}}(b)|}{|U|} = \frac{|\{1,2,3,4,5,6,7\}|}{|\{1,2,3,4,5,6,7\}|} = \frac{7}{7} = 1$$

$$\gamma_{T \cup \{bcd\}}(c) = \frac{|POS_{T \cup \{bcd\}}(c)|}{|U|} = \frac{|\{1,2,3,4,5,6,7\}|}{|\{1,2,3,4,5,6,7\}|} = \frac{7}{7} = 1$$

$$\gamma_{T \cup \{bcd\}}(d) = \frac{|POS_{T \cup \{bcd\}}(d)|}{|U|} = \frac{|\{1,2,3,4,5,6,7\}|}{|\{1,2,3,4,5,6,7\}|} = \frac{7}{7} = 1$$

$$\overline{\gamma_{T \cup \{bcd\}}}(y); \forall y \in C = 1$$

The subset {b, c, d} produces the dependency equal to '1' hence algorithm stops and outputs R = {b, c, d} as Reduct.

6.6   Summary

There are a number of feature selection algorithms employing Rough Set Theory for unsupervised datasets. RST theory is equally effective for unsupervised feature selection as in case of the supervised mod. In this chapter, we have presented a few of the most commonly refereed feature selection algorithms. The details of each algorithm along the description of pseudocode were also provided. Working examples are also explained in order to mention the exact working of the algorithms.

References

1. Velayutham C, Thangavel K (2011) Unsupervised quick reduct algorithm using rough set theory. J Electron Sci Technol 9(3):193–201
2. Velayutham C, Thangavel K (2011) Rough set based unsupervised feature selection using relative dependency measures. In: Digital Proceedings of UGC Sponsored National Conference on Emerging Computing Paradigms (2011)
3. Parthaláin NM, Jensen R (2010) Measures for unsupervised fuzzy-rough feature selection. Int J Hybrid Intell Syst 7(4):249–259
4. Inbarani HH, Nizar Banu PK (2012) Unsupervised hybrid PSO—relative reduct approach for feature reduction. In: 2012 International Conference on Pattern Recognition, Informatics and Medical Engineering (PRIME). IEEE
5. Nizar Banu PK, Inbarani HH (2012) Performance evaluation of hybridized rough set based unsupervised approaches for gene selection. Int J Comput Intell Inf 2(2):132–141

Chapter 7
Critical Analysis of Feature Selection Algorithms

So far in previous chapters, we have discussed details of various feature selection algorithms, both rough set based and non-rough set based, for supervised learning and unsupervised learning. In this chapter, we will provide an analysis of different RST-based feature selection algorithms. With explicit discussion on their results. Different experiments were performed to compare the performance of algorithms. We will focus on RST-based feature selection algorithms.

7.1 Pros and Cons of Feature Selection Techniques

Feature selection algorithms can be classified into three categories as follows:

- Filter methods,
- Wrapper methods,
- Embedded methods.

We have already provided a detailed discussion on all of these techniques. Here we will discuss some plus and negative points of all of these approaches.

7.1.1 Filter Methods

Filter methods perform feature selection independent of the underlying learning algorithm. Normally such methods rank feature according to some ranking criteria and select the best features. They ignore the impact of the feature on the learning algorithm. Here are some advantage and disadvantages of this approach:

© Springer Nature Singapore Pte Ltd. 2019
M. S. Raza and U. Qamar, *Understanding and Using Rough Set Based Feature Selection: Concepts, Techniques and Applications*, https://doi.org/10.1007/978-981-32-9166-9_7

Pros:

- Scalable to high dimensional datasets.
- Simple and fast.
- Independent of the underlying classification algorithm so can be used with any classification algorithm.

Cons:

As classification algorithm is ignored during feature ranking, so the selection of features may affect accuracy and performance of classification algorithm

- Ranks features independently and thus ignores dependencies among features.
- Normally univariate or low-variate.

7.1.2   Wrapper Methods

Perform feature selection based on the classification algorithm. Features are selected by using the feedback from the classifier. Following are some of the advantages and disadvantages of this approach:

Pros:

- Feature selection is performed by using the feedback of the classification algorithm.
- Enhances performance and accuracy of the classifier.
- Feature dependencies are also considered.

Cons:

- As compared to filter techniques, wrappers are highly prone to over fitting.
- Computationally expensive.

7.1.3   Embedded Methods

Perform feature selection as part of the learning procedure e.g. classification trees, learning machines etc.

Pros:

- As compared to wrapper methods they are less expensive.

Cons:

- Learning machine dependent.

7.2   Comparison Framework

A comparison framework was designed by authors in [1] to perform an analysis of different feature selection algorithms. The framework comprises of three components discussed below.

7.2.1   Percentage Decrease in Execution Time

Percentage decrease in execution time specifies the efficiency of an algorithm in terms of how fast it is and how much execution time it cuts down. For this purpose, system stopwatch was used, which after feeding the input was started and after getting the results was stopped. The formula to calculate the % decrease is as follows:

$$Percentage\,Decrease = 100 - \frac{E(1)}{E(2)} * 100$$

where E(1) is the execution time of one algorithm and E(2) is that of its competitor.

7.2.2   Memory Usage

Memory usage specifies the maximum amount of runtime memory taken by the algorithm to complete the task taken during its execution. Memory usage metric calculates memory by summing the size of each of the intermediate data structure used.

We have executed feature selection algorithms using 'Optidigits' data set from UCI [2]. Table 7.1 shows the results obtained.

Figure 7.1 shows a graph of comparison of execution time.

Table 7.1 Execution time of feature selection algorithms

PSO-QR time (m)	FSA time (m)	RS-IHS-QR time (m)	GA time (m)	IFSA time (m)	FSHE time (m)
6.52	2.21	0.46	0.40	1.22	12.24
46.11%	81.73%	96.19%	96.69%	89.91%	–

Fig. 7.1 Graph of execution
time of various FS algorithms

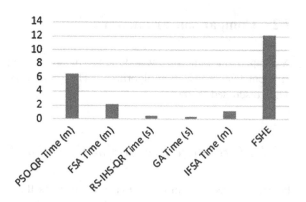

It can be noted that FSHE took maximum execution time. It is due to the reason that it first uses a heuristics approach and then applies exhaustive search on the result produced. The minimum time was taken by genetic algorithm. It gave a maximum decrease in execution time w.r.t. FSHE algorithm.

As far as memory is concerned, the same amount of memory was taken by all algorithms as the dataset used was same. Note that we only considered the major data structures to calculate memory and the intermediate/local variables were neglected.

7.3 Critical Analysis of Various Feature Selection Algorithms

Now we will present a critical analysis of various feature selection techniques. Strengths and weaknesses of each will be discussed.

7.3.1 Quick Reduct

QuickReduct [3] attempts to find feature subset without exhaustively generating all possible subsets. It is one of the most commonly referred algorithms, however, QuickReduct does not ensure optimal feature subset selection. During experimentation it was found that performance of QuickReduct depends on the distribution of attributes, e.g. in case of forward feature selection if attributes that result in higher degree of dependency are found at the end, the algorithm will take more time as compared to if the attributes are found in starting indexes. We explain this with the help of the following example. Consider the following dataset (Table 7.2).

In first iteration algorithm selects 'b' as part of reduct set, however in second iteration algorithm will pick at first iteration 'a, b' as reduct because the set {a, b}

Table 7.2 Sample dataset

U	a	b	c	D
X_1	0	0	0	x
X_2	1	1	1	y
X_3	1	1	0	y
X_4	2	0	1	x
X_5	0	2	0	z
X_6	2	2	0	y

Table 7.3 Sample dataset after changing attribute sequence

U	c	b	a	D
X_1	0	0	0	x
X_2	1	1	1	y
X_3	0	1	1	y
X_4	1	0	2	x
X_5	0	2	0	z
X_6	0	2	2	y

results in dependency equal to '1'. However, if attributes are distributed as follows (Table 7.3).

Then in second iteration, the algorithm will first combine 'b' with 'a' and then 'b' with 'c', hence taking more time. Thus performance of algorithm depends on distribution of attributes. Similarly, the number of attributes in Reduct set may also vary depending upon the distribution of attributes.

7.3.2 Rough Set Based Genetic Algorithm

In [4] authors present a genetic algorithm for feature subset selection. Genetic algorithms have been successfully applied to find feature subsets, however, these algorithms suffer following drawbacks:

- Randomness forms the core of genetic algorithms however, this may lead to losing important features containing more information than others, e.g. in a chromosome, there may be attributes having lower individual dependency but the cumulative dependency of '1' and thus the attributes represented through genes in such chromosome qualifying for the feature subset. However, the dataset may still have attributes having higher dependency (thus more information) as compared to these low dependency attributes (represented by chromosome).
- Genetic algorithms do not ensure the optimal feature subset selection as attributes are randomly selected, so the chromosome may contain redundant

attributes e.g. a chromosome may contain ten attributes out of which only three may be sufficient (providing dependency equal to that of entire attribute set).

- Evaluation of fitness function normally acts as a bottleneck for genetic algorithm. In context of RST-based feature selection, normally the evaluation function comprises of calculating dependency of decision class 'D' on set of conditional attributes encoded into the current chromosome. However, for large population size (more number of chromosomes) or larger datasets, calculating dependency is really a computationally expensive job seriously affecting the performance of the algorithm.
- Normally algorithm requires more number of chromosomes in a population (larger population size) and in some cases (as observed in experimentation), the algorithm was explicitly forced to stop after a specified number of iterations without producing the optimal results.
- The operators like crossover, mutation, etc. are important for the efficiency of the algorithm but there are no proper heuristics as to use which operator in which situation. Same is the case with the chromosome encoding scheme.
- The stopping criteria is not clear, e.g. the maximum number of iterations (in case of no optimal solution found) are manually specified and no guidelines regarding it.

7.3.3 PSO-QR

PSO-QR [5] is rough set based hybrid feature selection approach using swarm optimization combined with QuickReduct. The intention was to take advantage of both QuickReduct and PSO. The algorithm, however, inherits some limitations of both. Here we mention some of the limitations as follows:

- As with genetic algorithm, PSO-QR uses random particle generation, '1' represents the presence of an attribute and '0' represents absence. For example, following particle shows that first, third and fifth attributes will be considered and second and fourth will be missed.
  However, it suffers the same problem, i.e. particles may contain redundant attributes. So again the result produced is not optimal.
- It uses a rough set based dependency measure based on positive region, so not suitable for particles with large size of swarm with large number of particles.
- Furthermore, it uses QuickReduct as shown in the following step of the algorithm, which causes serious performance bottlenecks.
- No proper guidelines to select the values of inertia weights, furthermore, there are chances that algorithm traps in the local optima.
- There is no clear stopping criteria as well. Normally algorithm terminates after a specified number of iterations either the ideal solution has reached or not.

Fig. 7.2 Sample particle

| 1 | 0 | 1 | 0 | 1 |

Fig. 7.3 QuickReduct step in PSO-QR

$$\forall x \in (C\text{-}R)$$

$$\gamma_{R \cup (x)}(D) = \frac{|POS_{R \cup (x)}(D)|}{|U|}$$

- The method easily suffers from the partial optimism, which causes the less exact at the regulation of its speed and the direction [6].

It should, however, be noted that PSO does not involve the genetic operators like crossover, mutation, etc. However, particles change their velocity and position with respect to the global best particle. It means that less change as compared to chromosomes in the genetic algorithm (Figs. 7.2 and 7.3).

7.3.4   Incremental Feature Selection Algorithm (IFSA)

Incremental feature selection algorithm considers the dynamic dataset where attributes are added dynamically with the passage of time. One of the positive aspects of the algorithm is that it provides an explicit optimization step to remove the redundant attributes. The task is completed through measuring the significance of the attribute, i.e. dependency is calculated before and after removing attribute, if removal leads to decrease in dependency, it means the attribute is not redundant and should not be removed. Algorithm, however, suffers from the following limitations:

- Significance is calculated using the expression: $sig_2(a,B,D) = \gamma B \cup \{a\}$ $(D) - \gamma B(D)$, which means that we will have to calculate dependency twice, first with attribute 'a' and secondly by excluding the attribute. This results serious performance degradation in case of a large number of attributes.
- Algorithm converges to non-incremental feature selection algorithms in case of lesser number of features already known or more number of features added dynamically.

7.3.5   AFSA

Authors in [7] have presented a hybridized approach for feature selection using rough set theory and fish swarm logic. The algorithm uses swarm logic to iteratively optimize the solution to feature selection. Apart from the positive aspects of swarm logic AFSA suffers from following limitations [8]:

Table 7.4 PSO based algorithms

PSO algorithm	Description
Feature selection and parameter optimization of support vector machines based on modified artificial fish swarm algorithms [9]	Research proposes a modified AFSA (MAFSA) to improve feature selection and parameter optimization for support vector machine classifiers
Feature optimization based on the artificial fish swarm algorithm in intrusion detections [10]	Research proposes a method of optimization and simplification to network feature using artificial fish swarm algorithm in intrusion detection
Dataset reduction using improved fish swarm algorithm [11]	Research proposes a new intelligent swarm modeling approach that consists primarily of searching, swarming, and following behaviors
Evolving neural network classifiers and feature subset using artificial fish swarm [12]	Research presents the use of AFSA as a new tool which sets up a neural network (NN), adjusts its parameters, and performs feature reduction, all simultaneously
Feature selection for support vector machines base on modified artificial fish swarm algorithm [13]	Research proposes a modified version of artificial fish swarm algorithm to select the optimal feature subset to improve the classification accuracy for support vector machines

- Higher time complexity.
- Lower convergence speed.
- Lack of balance between global search and local search.
- Not use of the experiences of group members for the next moves.
- Furthermore, there are no guidelines for input parameters. Efforts are made from time to time to improve these limitations.

The Table 7.4 presents some variations on basic fish swarm.

7.3.6   Feature Selection Using Exhaustive and Heuristic Approach

Authors in [14] propose a new feature selection approach taking advantage of both heuristic and exhaustive search algorithms. The approach in this way can avoid the limitations of both categories of search. Exhaustive search is impossible and takes too much time thus cannot be used for large scale datasets, heuristics search, on the other hand, does not provide the optimal results. In [14] authors first performed the heuristic approach to perform feature selection. After this they performed an exhaustive search to remove the redundant features, thus ensuring the optimal feature subset.

The results were impressive as compared to bare use of exhaustive search, however, it suffers the from following limitations:

- Algorithm experiences too much performance degradation as compared to other feature selection algorithms employing only one strategy.
- The limitations of the heuristic search algorithms like no explicit guidelines for input, no proper stopping criteria, etc. could not be removed.

7.3.7 Feature Selection Using Random Feature Vectors

In [15], the authors have presented a feature selection approach based on hit and trial method. Research is based on randomly generating feature subsets until we get the one having maximum dependency, i.e. equal to that of the entire conditional attribute set. It then performs optimization step to remove the redundant features.

The proposed algorithm is better than [14] as it calculates feature vectors without using complex operators like GA, PSO, AFS, etc. However, the algorithm suffers from the following limitations:

- No explicit stopping criteria in case if hit and trial do not generate any ideal solution.
- Optimizing the resultant feature vector requires calculating dependency twice, once by including a feature and then by excluding it which may result in a performance bottleneck.

7.4 Summary

In this chapter, we have performed critical analysis of a few of the most commonly used algorithms. The intention was to elaborate on the strengths and weaknesses of each approach. For this purpose firstly experimentation was conducted a benchmark dataset and results were discussed. Then few of the algorithms were discussed in depth along with their strengths and weakness. This would give the research community a direction to further investigate these algorithms and overcome their limitations.

References

1. Raza MS, Qamar U (2016) An incremental dependency calculation technique for feature selection using rough sets. Inf Sci 343:41–65
2. Lichman M (2013) UCI machine learning repository. University of California, School of information and computer science, Irvine, CA. http://archive.ics.uci.edu/ml. Last accessed 30 March 2017

3. Jensen R, Shen Q (2008) Computational intelligence and feature selection: rough and fuzzy approaches. Wiley
4. Zuhtuogullari K, Allahvardi N, Arikan N (2013) Genetic algorithm and rough sets based hybrid approach for reduction of the input attributes in medical systems. Int J Innov Comput Inf Control 9:3015–3037
5. Inbarani HH, Azar AT, Jothi G (2014) Supervised hybrid feature selection based on PSO and rough sets for medical diagnosis. Comput Methods Programs Biomed 113(1):175–185
6. Bai Q (2010) Analysis of particle swarm optimization algorithm. Comput Inf Sci 3(1):180
7. Chen Y, Zhu Q, Xu H (2015) Finding rough set reducts with fish swarm algorithm. Knowl Based Syst 81:22–29
8. Neshat M, et al (2014) Artificial fish swarm algorithm: a survey of the state-of-the-art, hybridization, combinatorial and indicative applications. Artif Intell Rev 42(4):965–997
9. Lin KC, Chen SY, Hung JC (2015) Feature selection and parameter optimization of support vector machines based on modified artificial fish swarm algorithms. Math Probl Eng (2015)
10. Liu T, et al (2009) Feature optimization based on artificial fish-swarm algorithm in intrusion detections. In: International conference on networks security, wireless communications and trusted computing. NSWCTC'09. vol 1, IEEE
11. Manjupriankal M, et al (2016) Dataset reduction using improved fish swarm algorithm. Int J Eng Sci Comput 6(4):3997–4000
12. Zhang M, et al (2006) Evolving neural network classifiers and feature subset using artificial fish swarm. In: Proceedings of the 2006 IEEE international conference on mechatronics and automation. IEEE
13. Lin KC, Chen SY, Hung JC (2015) Feature selection for support vector machines base on modified artificial fish swarm algorithm. In: Ubiquitous computing application and wireless sensor. Springer, Berlin, pp 297–304
14. Raza MS, Qamar U (2016) A hybrid feature selection approach based on heuristic and exhaustive algorithms using rough set theory. In: Proceedings of the international conference on internet of things and cloud computing. ACM
15. Raza MS, Qamar U (2016) A rough set based feature selection approach using random feature vectors. 2016 international conference on frontiers of information technology (FIT). IEEE

Chapter 8
Dominance-Based Rough Set Approach

Dominance-based Rough Set Approach (DRSA) is an extension of conventional Rough Set Theory. The conventional approach does not take into consideration the dominance relation among the objects while considering the universe. DRSA, on the other hand, considers this relation as well and thus takes RST a step ahead. In this chapter, we will consider some preliminaries of DRSA.

8.1 Introduction

Data analysis involves various tasks, e.g. classification, clustering, rule extraction, etc. If we talk about classification, we should consider some prior knowledge that may include [1]:

(i) Domain of attributes, i.e. the set of possible values each attribute can take.
(ii) Division of attributes, i.e. set of conditional and decision attributes and a relation between them.
(iii) A preference order in values of these attributes. The classification should preserve the preference order.

Classification, conventionally, includes both the item (i) and (ii) and all the tools almost sport the same. However, an important aspect that is missed is the preference order in the domain values of the attributes, i.e. item (iii). Objects are classified on the basis of the domain values of attributes and classes don't have any preference order. In practice, we come across many situations where attributes not only maintain order but have a preference order, i.e. certain values are preferred over others. For example consider that in a class, final grades are calculated on the basis of marks in Chemistry and Physics. Here the decision attribute, i.e. Final-Grade should have a preference order which means that the Final-Grade 'Excellent' is preferred over 'Very good', similarly, the Final-Grade 'Very good' is preferred over grade 'Good' and so on. Now if two students 'A' and 'B' get the same score in

© Springer Nature Singapore Pte Ltd. 2019
M. S. Raza and U. Qamar, *Understanding and Using Rough Set Based Feature Selection: Concepts, Techniques and Applications*, https://doi.org/10.1007/978-981-32-9166-9_8

Chemistry but Student 'A' gets higher score in 'Physics' than student 'B' then student 'A' should be assigned to preferred Final-Grades class. If this is not the case then it can be concluded that there is some inconsistency in getting the sample for example noise or missing values, etc. Handling these inconsistencies is a core issue in knowledge discovery. In real-life applications, these consistencies cannot simply be considered as noise or errors and normalized by some operator. Consideration of preference order becomes critical.

The attributes having preference ordered domains are called criteria. Both conditional and decision attributes may have preference ordered domains and hence may be called criteria.

Many tools and techniques have been proposed in the literature to perform data analysis and several of its tasks, Rough Set Theory, Decision Theoretic Rough Sets, Fuzzy Theory, etc., are few to name. However, all of them fail to consider the preference order which may consequence in inconsistent results. Dominance-Based Rough Set Approach is an extension to conventional Rough Set Theory (RST), which considers the preference orders as well. It generalizes the classical RST by substituting indiscernibility relation with dominance principals.

8.2   Dominance-Based Rough Set Approach

Dominance-based rough set approach proposed by Greco, Matarazzo and Słowiński [2–4] as an extension to conventional RST. The extension includes the ordinal evaluations of objects in a dataset along with monotonic relationships between these evaluations. However, it should be noted that DRSA can be used for data analysis of non-ordinal problems as well [5]. Right from its emergence, it has been used in many domains, e.g. in manufacturing industry [6], finance [7, 8], project selection [9], data mining [10, 11], etc. In this section, we will discuss some of the core preliminaries of DSRA.

8.2.1   Decision Table

In conventional RST, a decision system is a finite set of objects called universe U, where each object is characterized by a set of conditional attributes C and decision attributes D. Mathematically,

$$\alpha = (U, C \cup D)$$

In the context of DSRA, a decision table is a four tuple mathematically represented as

Table 8.1 Decision system

U	Physics	Chemistry	Final-grade
X_1	A	B	Very good
X_2	A	A	Excellent
X_3	B	C	Good
X_4	A	B	Good
X_5	B	A	Very good
X_6	A	B	Very good
X_7	C	B	Good

$$\alpha = (U, Q, V, f)$$

Here U represents a finite set of objects, Q is a finite set of criteria, i.e. the attributes having ordinal scale based domain. Here, $V = \cup_{q \in Q} V_q$ where V_q is the value set of criteria q. f represents a function of the form f(x,q) which assigns a particular value V_q to an object x for attribute q. Here $Q = (C \cup D)$, i.e. both conditional attributes and decision attribute(s) are called included. Table 8.1 shows the sample decision system.

Here, universe comprises of seven objects {X1, X2, X3..., X7}. Conditional criteria include {Physics, Chemistry} and decision criteria is {Final-Grade}.

8.2.2 Dominance

Dominance specifies the preference order. For a criteria $P \subseteq C$, an object x dominates object y if x is better than y on every criterion from P, i.e. . In simple words, equal to $x_q \geq y_q$. It will be specified as 'x dominates y' and mathematically:

$$D_P^-(x) = \{y \in \cup : xD_Py\}$$

i.e. a set of objects dominated by x by considering the information in $P \subseteq C$.
Similarly, we can specify the set of objects dominating x as follows:

$$D_P^+(x) = \{y \in \cup : yD_Px\}$$

If we consider the Table 8.1 and object X_3 as our origin and P = {Physics} then:

$$D_P^-(x) = \{X_5, X_7\}$$

Similarly,

$$D_P^+(x) = \{X_1, X_2, X_5, X_6\}$$

8.2.3   Decision Classes and Class Unions

In conventional RST, the decision attribute provides a partition of the universe in finite number of decision classes. Similarly, in the case of DSRA, decision attribute divides the universe in finite decision classes $Cl = \{Cl_1, Cl_2, Cl_3, \ldots, X_m\}$. Note that each object belongs to one and only one decision class.

However, in contrast with RST, in DSRA the decision classes are assumed to be preference ordered. So, for r, s = {1, 2, 3..., m}, object belonging to Cl_r is preferred over the object belonging to Cl_s for r > s. So, instead of simple approximation as in case of RST, the approximation in DSRA is upward unions and downward unions of classes.

$$Cl_t^{\geq}(x) = \cup_{s \geq t} Cl_s \quad t = 1, \ldots n.$$

$$Cl_t^{\leq}(x) = \cup_{s \leq t} Cl_s \quad t = 1, \ldots n.$$

Here, $Cl_t^{\geq}(x)$ specifies the set of objects belonging to Cl_t or a more preferred class. $Cl_t^{\geq}(x)$, on the other hand specifies the set of objects belonging to Cl_t or to a less preferred class.

From Table 8.1, the decision attribute 'Final-Grades' has three decision classes which are 'Excellent', 'Very Good' and 'Good'. 'Excellent' is preferred over 'Very Good' which is preferred over 'Good'. Considering their indexes as Excellent = 3, Very Good = 2 and Good = 1 then for t = 2:

$$Cl_t^{\geq}(x) = \{X_1, X_2, X_5, X_6\}$$

i.e. $Cl_t^{\geq}(x)$ specifies the set of objects either belonging to the class 'Very good' or more preferred class i.e. 'Excellent'.

Similarly

$$Cl_t^{\leq}(x) = \{X_1, X_3, X_4, X_5, X_6, X_7\}$$

i.e.

$Cl_t^{\leq}(x)$ specifies the set of objects belonging to class 'Very good' or less preferred, i.e. 'Good'.

For t = 1:

$$Cl_t^{\geq}(x) = \{X_1, X_2, X_3, X_4, X_5, X_6, X_7\}$$

$$Cl_t^{\leq}(x) = \{X_3, X_4, X_7\}$$

8.2.4 Lower Approximations

Lower approximation in RST defines the set of objects that with certainty will belong to a decision class with respect to the given attributes. In DSRA, given that $P \subseteq C$ P-lower approximation of $Cl_t^{\geq}(x)$ specifies all the objects that will, with certainty, belong to $Cl_t^{\geq}(x)$. Similarly, the P-lower approximation of $Cl_t^{\leq}(x)$ will contain all the objects that, with certainty, will belong to $Cl_t^{\leq}(x)$.

Mathematically,

$$\underline{P}(Cl_t^{\geq}) = \{x \in U : D_P^+(x) \subseteq Cl_t^{\geq}\}$$

For $Cl_t^{\leq}(x)$:

$$\underline{P}(Cl_t^{\leq}) = \{x \in U : D_P^-(x) \subseteq Cl_t^{\leq}\}$$

Calculating comprises of three steps explained here with the help of the sample data set given in Table 8.1. We will demonstrate the process for $\underline{P}(Cl_t^{\geq})$ below with step-by-step explanation of example.

Step-1: In the first step, we calculate the objects belonging to the union of classes $Cl_t^{\geq}(x)$. This is just like calculating equivalence class structure using decision attribute in conventional RST. In our example we will calculate lower approximation $\underline{P}(Cl_t^{\geq})$ for t = 2.

$$Cl_t^{\geq} = \{X1, X2, X5, X6\}$$

Step-2: In the second step, we calculate $D_p^+(x)$ for each object identified in Step 1. In our case

For X1: $D_P^+(x1) = \{X1, X2, X4, X6\}$
For X2: $D_P^+(X2) = \{X2\}$
For X5: $D_P^+(X5) = \{X2, X5\}$
For X6: $D_P^+(X6) = \{X1, X2, X4, X6\}$

It is evident from the above example that we have to calculate $D_p^+(x)$ for each individual object which is a computationally expensive step as it requires multiple dataset passes which can be computationally expensive in case of dataset beyond smaller size.

Step-3: In the third step, we actually calculate lower approximation. The sets identified in step-2 that are a subset of the sets identified in Step 1 become part of the lower approximation. These are the objects about which we can, with certainty, conclude that they belong to the union of class t or preferred. So,

$$\underline{P}(Cl_t^{\geq}) = \{X2, X5\}$$

Note that we have to follow the same three steps in calculating the lower approximation for $P(Cl_t^{\leq})$

8.2.5 Upper Approximations

In conventional RST based approach, the upper approximation defines that set of objects that may possibly belong to the concept X. In DSRA, for $P \subseteq C$ the P-upper approximation of $Cl_t^{\geq}(x)$ defines the set of objects that may possibly belong to the union of classes $Cl_t^{\geq}(x)$. Similarly, the P-upper approximation of $Cl_t^{\leq}(x)$ defines the set of objects that may possibly belong to the union of classes $Cl_t^{\leq}(x)$.

Mathematically,

$$P(Cl_t^{\geq}) = \{x \in U : D_P^-(x) \cap Cl_t^{\geq} \neq \emptyset\}$$

For $Cl_t^{\leq}$:

$$P(Cl_t^{\leq}) = \{x \in U : D_P^+(x) \cap Cl_t^{\leq} \neq \emptyset\}$$

i.e. we cannot with certainty conclude that the object will belong to the union of $Cl_t^{\geq}(x)$. Same is the case for $\bar{P}(Cl_t^{\leq})$.

Calculating upper approximation also comprises of three steps. Now we will explain with the help of an example about how to calculate upper approximation.

We will use Table 8.1 given above and will calculate p-upper approximation $\bar{P}(Cl_t^{\leq})$ for t = 2.

Step-1: Just like P-lower approximation, in calculating P-upper approximation, the first step is to calculate all the objects belonging to the union of classes $Cl_t^{\geq}$. In our case:

$$Cl_t^{\geq} = \{X1, X2, X5, X6\}$$

Step-2: In the second step, we calculate $D_P^-(x1)$ for each object belonging to the set identified in step-1. In our case

ForX1: $D_P^-(x1) = \{X1, X3, X4, X6, X7\}$
For X2: $D_P^-(X2) = \{X2\}$
For X5: $D_P^-(X5) = \{X3, X5, X7\}$
For X6: $D_P^-(X6) = \{X1, X3, X4, X6, X7\}$

It should be noted that this step significantly degrades the performance as for each object we need to find objects dominating it. This requires a complete traversal of the dataset for each object. So, as we have four objects in $Cl_t^{\geq}$, so we will have to perform four traversals of the dataset.

Step-3: Finally, in step-3 we determine the objects belonging to the P-upper approximation. This requires identifying the objects in subsets (identified in Step-2) that have non-empty interaction with the set identified in Step-1.

In our case, all $D_P^-(X1)$, $D_P^-(X2)$, $D_P^-(X5)$ and $D_P^-(X6)$ have non-empty interaction. So, $\bar{P}(Cl_t^\geq)$ for t = 2 will be

$$\bar{P}(Cl_t^\geq) = X1, X2, X3, X4, X5, X6, X7,$$

The same three steps will be performed for $\bar{P}(Cl_t^\geq)$. All of the DSRA-based algorithms use this approach which affects the performance of the algorithm and consequently, these algorithms cannot be used for datasets beyond smaller size without significant degrading the performance.

Pseudocode for calculating approximations:

DRSA almost uses the same method to calculate these approximations, however, DRSA additionally considers dominance relation as well. So, it is equally challenging in DRSA. To calculate these approximations using the conventional method requires three steps. In the first step we calculate $Cl_t^\geq$ or $Cl_t^\leq$ structure depending on which approximation we need to calculate.

For example, Fig. 8.1 shows the pseudo code to calculate $Cl_t^\geq$ in case of $\bar{P}(Cl_t^\geq)$.

In the provided pseudocode, Cl_i represents the decision class of the ith object in dataset while Cl_t represents the decision class for which we need to calculate $P(Cl_t^\geq)$. Here we need to traverse the complete dataset for calculating $P(Cl_t^\geq)$.

Now in second step, we calculate $D_P^+(x)$ which comprises of all the objects greater than or equal to each object in $Cl_t^\geq(x)$. This means that if $Cl_t^\geq(x)$ comprises of five objects we need to traverse the dataset five times to calculate $D_P^+(x)$ for each object. Having larger datasets means more number of objects in $Cl_t^\geq(x)$ and thus more number of dataset traversals, which significantly effects the performance of the algorithm. The pseudo code for this step is given in Fig. 8.2.

Here X_j represents jth object in the dataset and X_{it} represents ith object in $Cl_t^\geq$.

Now, finally we calculate $P(Cl_t^\geq)$ which comprises of the objects (identified in second step) that are subset of objects identified in first step. Figure 8.3 shows the pseudocode of this step.

Here, $D_P^+(X_{ji})$ represents the jth object in the set comprising of all objects greater than X_i in $Cl_t^\geq$ and $Cl_{kt}^\geq$ represents the kth object in $Cl_t^\geq$.

$$
\begin{array}{l}
\forall i \in U \\
\quad \text{If } Cl_i \geq Cl_t \\
\quad\quad Cl_t^\geq = Cl_t^\geq \cup Cl_i
\end{array}
$$

Fig. 8.1 Pseudocode of Step 1 for calculating the lower approximation

$$\forall i \in Cl_t^{\geq}$$
$$\forall j \in U$$
$$If\, X_j \geq X_{it}$$
$$D_P^+(X_i) = D_P^+(X_i) \cup X_j$$

Fig. 8.2 Pseudocode of Step 1 for calculating the lower approximation

$$\forall i \in Cl_t^{\geq}$$
$$\forall j \in D_P^+(X_i)$$
$$\forall k \in Cl_t^{\geq}$$
$$Calculate\, D_P^+(X_{ji}) \subseteq Cl_{kt}^{\geq}$$

Fig. 8.3 Pseudocode of Step 1 for calculating the lower approximation

It is clear that the conventional approach poses serious challenges to performance of the algorithms using these approaches when it comes to larger datasets. We, therefore need a more efficient method to calculate both lower and upper approximations.

8.3   Some DRSA-Based Approaches

Since its inception, DRSA has been used in many domains for different tasks. Here we will discuss few DRSA-based algorithms taken from literature.

In [12], the authors have used improved DSRA for classification of medical data. DSRA is used for ordinal attributes, the proposed technique is used for nominal ones. It suggests the decision table to determine dominance relation, the improved DSRA is applied to determine the lower and upper approximations in the entire dataset. Finally, the attribute reduction technique is applied to find the reduced number of the attribute for classification.

The proposed technique comprises five steps. In the first step, they construct the decision table to apply DSRA, then based on this decision table, the lower and upper approximations are calculated using the conventional methods. In the third step, the boundary values and dependency are calculated. Fourth, the Reduct and core are found out to apply feature selection and finally the rule generation step for classification.

Figure 8.4 shows its five-step methodology.

In [13], authors have used DSRA to predict the number of students likely to drop out from the Massive Open Online Course (MOOCS) course next week using the historic data of the previous week. The proposed approach proposes two classes of students Cl1 which specifies the 'At-risk Learns' and Cl2 which specifies 'Active Learners'.

Step 1: Construction of decision table to apply dominance rough set.
Step 2: Based on decision table find lower and upper approximation using Equations:

$$\underline{P}(Cl_t^{\geq}) = \{x \in U: D_P^+(x) \subseteq Cl_t^{\geq}\}$$
$$\underline{P}(Cl_t^{\leq}) = \{x \in U: D_P^-(x) \subseteq Cl_t^{\leq}\}$$
$$\overline{P}(Cl_t^{\geq}) = \{x \in U: D_P^-(x) \cap Cl_t^{\geq} \neq \emptyset\}$$
$$\overline{P}(Cl_t^{\leq}) = \{x \in U: D_P^+(x) \cap Cl_t^{\leq} \neq \emptyset\}$$

Step 3: Find boundary values and dependency.
Step 4: Find reduct set and core set to apply feature selection.
Step 5: Apply rule generation for classification.

Fig. 8.4 Five-step methodology for the improved dominance-based rough set (IDRSA) taken from [12]

The proposed approach is a two-step method, the first step inferences a preference mode, while the second phase classifies the students in the above-mentioned classes. The first step itself consists of three steps. The first step identifies the learning examples of learners. The second step constructs the coherent criteria family for learners profile characterization and finally, the third step is to infer a preference model resulting in a set of decision rules.

Figure 8.5 shows the two-phase methodology of the proposed approach.

In [14], the authors have proposed DSRA based approach for predicting customer behaviour in airline companies. This can help managers attain new customers and retain high valued customers. A set of rules is derived from a large sample of international airline customers, and its predictive ability is evaluated. Results have shown the effectiveness of the approach. In [15], authors have proposed a new approach for finding the Reducts in DSRA. They have investigated the attribute reduction in DSRA along with introducing class-based Reducts and their relations with previous Reducts.

Class-based Reducts are of three kinds. The first kind of Reducts, called L-Reduct, preserves the lower approximations of decision classes, the second kind Reduct, called U-Reduct, preserves the upper approximations of decision classes, and the third kind of Reduct, called B-Reduct, preserves the boundary regions of decision classes. They also show that all kinds of Reducts can be enumerated comprehensively based on two discernibility matrices associated with generalized decisions.

There are many other approaches [16–19] using DSRA for different purposes. However, all the approaches discussed so far use static datasets, i.e. they are based on the concept that underlying data is complete and no further data will be added at

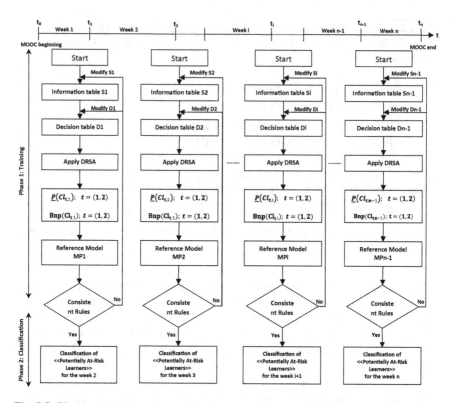

Fig. 8.5 Weekly prediction method of 'At-risk learners' based on DRSA was taken from [13]

runtime but there are many incremental approaches [20–23]. These approaches consider dynamic information systems to calculate approximations. In dynamic information systems, information keeps on adding with the passage of time. So, once approximation is calculated, we need to update it regularly as new information is added in the dataset. Figures 8.6 and 8.7 show the algorithms taken from [20] using both non-incremental and incremental approach for updating approximations of DRSA.

In [24], Shen and Tzeng, have used Dominance-based Rough Set Approach for prediction purpose. They find core attributes in edicion rules which are further processed by an integrated multiple criteria decision-making method to make the selection and to devise improvement plans. By using the VIKOR method and the influential weights of DANP, decision maker may be planned to reduce gap of each criterion for achieving the aspired level. The retrieved attributes (i.e. criteria) are used to collect the knowledge of domain experts for selection and improvement.

```
Input:
A decision system.
Output:
begin
for i = 1 → |U| do for j = 1 → |U| do Øᴾ(i, j) ← 0;
        for i = 1 → |U| do
                for j = 1 → |U| do
                        for s = 1 → |P| do
                                if f_{t+1}(x_i − a_s) ≥ f_{t+1}(x_j − a_s) then Øᴾ(i, j) ← Øᴾ(i, j) + 1;
                        end
                end
        end
        for i = 1 → |U| do
                Compute D_p^+(x_i) and D_p^-(x_i);
        end
        for n = 2 → m do
                Compute Cl_n^≥ and Cl_{n-1}^≤;
                Compute P̲(Cl_n^≥), P̄(Cl_n^≥), P̲(Cl_{n-1}^≤) and P̄(Cl_{n-1}^≤);
        end
        output the results
end
```

Fig. 8.6 A non-incremental algorithm for computing approximations of DRSA taken from [20]

```
input:
(1) The dominance matrix Rᴾ , the P-dominating and P-dominating sets at time t;
(2) The approximation of DRSA at time t;
(3) The attribute values varied at time t+1.
output:
The approximation of DRSA at time t+1.
begin
        for i = 1 → |U| do for j = 1 → |U| do Δ Øᴾ(i, j) ← 0;
        forall the (k, s) ∈ VV do  //(k, s) is an element of the set VV, where k is the index of the object
        whose value with respect to attribute a_s was changed during the dynamic process.
                for i = 1 → |U| do
                        if f_{t+1}(x_i, a_s) ≥ f_{t+1}(x_k, a_s) ∧ r_t^{a_s}(i, k) = 0 then Δ Øᴾ(i, k) ← Δ Øᴾ(i, k) + 1;
                        if f_{t+1}(x_i, a_s) < f_{t+1}(x_k, a_s) ∧ r_t^{a_s}(i, k) = 1 then Δ Øᴾ(i, k) ← Δ Øᴾ(i, k) − 1;
                        if f_{t+1}(x_i, a_s) ≥ f_{t+1}(x_k, a_s) ∧ r_t^{a_s}(k, i) = 0 then Δ Øᴾ(k, i) ← Δ Øᴾ(k, i) + 1;
                        if f_{t+1}(x_k, a_s) ≥ f_{t+1}(x_i, a_s) ∧ r_t^{a_s}(k, i) = 1 then Δ Øᴾ(k, i) ← Δ Øᴾ(k, i) − 1;
                end
        end
        for i = 1 → |U| do
                Compute Δ^+ D_p^+(x_i), Δ^- D_p^+(x_i), Δ^+ D_p^-(x_i) and Δ^- D_p^-(x_i);
                D_p^+(x_i)_{t+1} ← D_p^+(x_i)_t ∪ Δ^+ D_p^+(x_i) − Δ^- D_p^+(x_i);
                D_p^-(x_i)_{t+1} ← D_p^-(x_i)_t ∪ Δ^+ D_p^-(x_i) − Δ^- D_p^-(x_i);
        end
        for n = 2 → m do
                Compute Δ^+ P̲(Cl_n^≥), Δ^- P̲(Cl_n^≥), Δ^+ P̄(Cl_n^≥) and Δ^- P̄(Cl_n^≥);
                P̲(Cl_n^≥)_{t+1} ← P̲(Cl_n^≥)_t ∪ Δ^+ P̲(Cl_n^≥) − Δ^- P̲(Cl_n^≥);
                P̄(Cl_n^≥)_{t+1} ← P̄(Cl_n^≥)_t ∪ Δ^+ P̄(Cl_n^≥) − Δ^- P̄(Cl_n^≥);
                P̲(Cl_{n-1}^≤)_{t+1} ← P̲(Cl_{n-1}^≤)_t ∪ Δ^- P̄(Cl_n^≥) − Δ^+ P̄(Cl_n^≥);
                P̄(Cl_{n-1}^≤)_{t+1} ← P̄(Cl_{n-1}^≤)_t ∪ Δ^- P̲(Cl_n^≥) − Δ^+ P̲(Cl_n^≥);
        end
        output the results;
end
```

Fig. 8.7 An incremental algorithm for dynamically updating approximations of DRSA taken from [20]

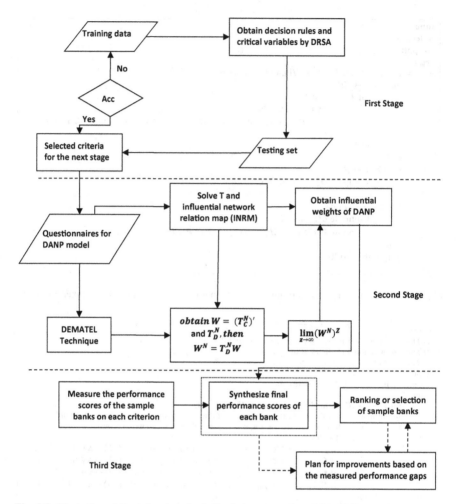

Fig. 8.8 Illustration of the infused methods for their proposed model in [24]

Figure 8.8 provides an illustration of the infused methods for their proposed model.

In [16], Du and Hu, the authors present attribute reduction approaches for incomplete information systems. A new kind of dominance relation, named the characteristic-based dominance relation, to incomplete ordered information systems, is introduced. The research also addresses the discernibility matrix and the discernibility function to compute all (relative) Reducts in incomplete ordered information systems (consistent incomplete ordered decision tables). To reduce the computational burden, a heuristic algorithm with polynomial time complexity for finding a single (relative) Reduct is designed by using the inner and outer significance measures of each criterion candidate.

Input: an incomplete ordered information system S $= (U, AT, V, f), A \subseteq AT$.
Output: all A-dominating sets of objects x from U, namely, $D_A^{\Diamond+}(x), \forall x \in U$.
1: **for each** $x, y \in U$ **do**
2:      set $m \leftarrow 0$;
3:      **for each** $a \in A$ **do**
4:      **if** $(f(y, a) =?$ **or** $f(y, a) =*$ **or** $f(x, a) =*$ **or** $f(y, a) \geq f(x, a))$ **then**
5:        $m \leftarrow m + 1$;
6:      **end if**
7:      **end for**
8:      **if** $m = |A|$ **then**
9:      $D_A^{\Diamond+}(x) \leftarrow D_A^{\Diamond+}(x) U y$;
10:     **end if**
11:     **return** $D_A^{\Diamond+}(x)$;
12: **end for**

Fig. 8.9 Algorithm of computation for all A-dominating sets

Input: A-dominating/dominated sets with respect to all objects from $U, U/D_d^+ = \{Cl_t^{\geq}\}$.
Output: all lower and upper approximations of $Cl_t^{\geq}$ with respect to A.
1:      **for each** $Cl_t^{\geq} \in U/D_d^+$ **do**
2:      **for each** $x \in U$ **do**
3:      **if** $D_A^{\Diamond+}(x) \cap Cl_t^{\geq} = D_A^{\Diamond+}(x)$ **then**
4:      $\underline{A}(Cl_t^{\geq}) \leftarrow \underline{A}(Cl_t^{\geq}) U x$;
5:      **end if**
6:      **if** $D_A^{\Diamond-}(x) \cap Cl_t^{\geq} \neq 0$ **then**
7:      $\overline{A}(Cl_t^{\geq}) \leftarrow \overline{A}(Cl_t^{\geq}) U x$;
8:      **end if**
9:      **end for**
10:     $Bn_A(Cl_t^{\geq}) \leftarrow \overline{A}(Cl_t^{\geq}) - \underline{A}(Cl_t^{\geq})$;
11:     **return** $\underline{A}(Cl_t^{\geq}), \overline{A}(Cl_t^{\geq}), Bn_A(Cl_t^{\geq})$;
12: **end for**

Fig. 8.10 Algorithm of computation for rough approximations

Figures 8.9, 8.10, 8.11 and 8.12 show pseudocode of the proposed algorithms taken from [16].

In [25], Maciag et al., refine their original procedure developed to personalize the interfaces for online shopping tools. Originally authors used Rough Set Theory, however, in [25] they replaced it with Dominance-Based Rough Set Approach and the two phases of their original procedure were analytically and empirically evaluated using results obtained when using DRSA. In the first phase, classification accuracy was slightly higher in the case of CRSA. In the second phase, however, they found that the original procedure could be further developed using the newly obtained information for additional design enhancements with results obtained from DRSA. The authors also discussed a possible third phase to the original procedure, incorporating results obtained from DRSA analyses to highlight products and features that target cross-cluster similarities between product feature values and consumer preferences.

Input: an incomplete ordered information system $S = (U, AT, V, f)$.
Output: all reducts of S.
1: for each $x, y \in U$ do // compute the discernibility matrix D
2:      $D(x, y) \leftarrow \emptyset$;
3:      for each $a \in AT$ do
4:      $if (f(x, a) < f(y, a)$ or $(f(y, a) =?$ and $f(x, a)$ is specified$))$ then
5:      $D(x, y) \leftarrow D(x, y) \cup a$; // put a into $D(x, y)$
6:      end if
7:              end for
8:      end for
9:      set $F_{\wedge(\vee)} \leftarrow 1$;
10:     for each $x, y \in U$ do
11:     if $D(x, y) \neq \emptyset$; then
12:     $F_{xy} = \vee \{a | a \in D(x, y)\}$; // disjunction expression of $D(x, y)$
13:     end if
14:     $F_{\wedge(\vee)} = F_{\wedge(\vee)} \wedge F_{xy}$; // discernibility function of S
15:     end for
16:     $F_{\vee(\wedge)} \leftarrow F_{\wedge(\vee)}$; // convert the conjunctive normal form $F_{\wedge(\vee)}$ to minimal disjunctive
normal form $F_{\vee(\wedge)}$
17:     return $RED \leftarrow \{red | red \in F_{\vee(\wedge)}\}$; // RED the set of all reducts

Fig. 8.11 Discernibility matrix method for computing all Reducts in an IOIS

Input: an incomplete ordered information system $S = (U, AT, V, f)$.
Output: a reduct of S.
1:      set $B \leftarrow \emptyset$; //initialize B, a reduct of S
2:      for each $a \in AT$ do
3:      compute $sig^{\geq}_{inner}(a, AT) = \frac{\sum_{x \in U} |D^{\Diamond +}_{AT-\{a\}}(x)|}{|U|^2} - \frac{\sum_{x \in U} |D^{\Diamond +}_{AT}(x)|}{|U|^2}$;
4:      $sig^{\geq}_{inner}(a, AT) > 0$ then // a is indispensable for AT
5:      $B \leftarrow B \cup a$;
6:      end if
7:      end for // B the core of S
8:      for each $a \in AT - B$ do
9:      compute $sig^{\geq}_{outter}(a, B) = \frac{\sum_{x \in U} |D^{\Diamond +}_{B}(x)|}{|U|^2} - \frac{\sum_{x \in U} |D^{\Diamond +}_{B \cup (a)}(x)|}{|U|^2}$;
10:     end for
11:     select an a which satisfies $sig^{\geq}_{outter}(a, B) = \max_{b \in AT-B} sig^{\geq}_{outter}(b, B)$;
12:     if $sig^{\geq}_{outter}(a, B) = 0$ then
13:     go to step 16;
14:     else $B \leftarrow B \cup a$;
15:     end if
16:     if $\theta^{\geq}_{B} = \theta^{\geq}_{AT}$ then // check the stopping criterion
17:     delete the redundant element in B;
18:     return B;
19:     else go to step 8;
20:     end if

Fig. 8.12 Heuristic algorithm of computation for the acquisition of a Reduct in an IOIS

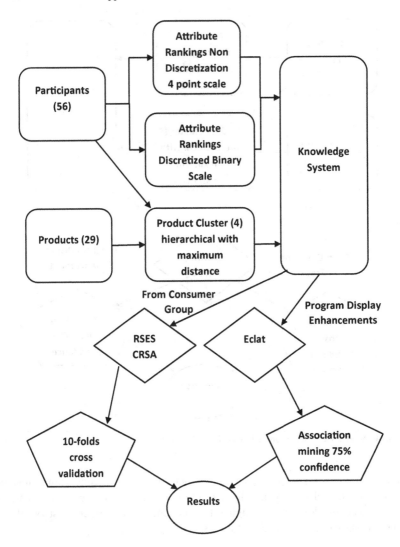

Fig. 8.13 Model view of the procedure using CRSA [25]

Figures 8.13 and 8.14 show the model view of the author's evaluations using CSRA and DSRA.

In [26], Zifu et al., use dominance rough set approach (DRSA) to construct the classification model for the judgment of emergency communication. In their model, they propose a classification index system of emergency communication using the method of expert interview first and then use DRSA to complete data sample, Reduct attribute, and extract the preference decision rules of the emergency

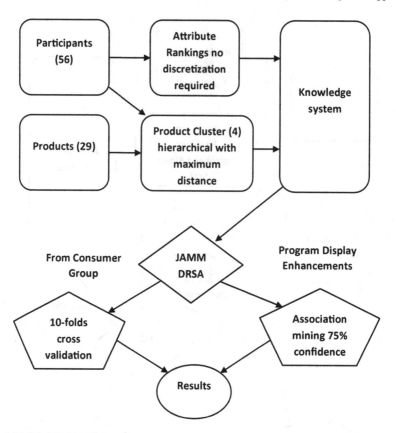

Fig. 8.14 Model view of the procedure using DRSA [25]

communication classification. The construction of emergency communication classification model was mainly divided into two phases: index extraction and data mining based on dominance rough set approach; the construction steps of their model are shown in Fig. 8.15 as follows.

8.4   Summary

In this chapter, we discussed DRSA in detail. Starting with a basic introduction, few preliminaries were presented with step-by-step examples along with pseudocode of approximations. Then we discussed some state of the art approaches from literature using DRSA. This chapter was intended to provide a strong base for DRSA and in the upcoming chapter, we will be discussing VBA code for calculating dominance and approximations.

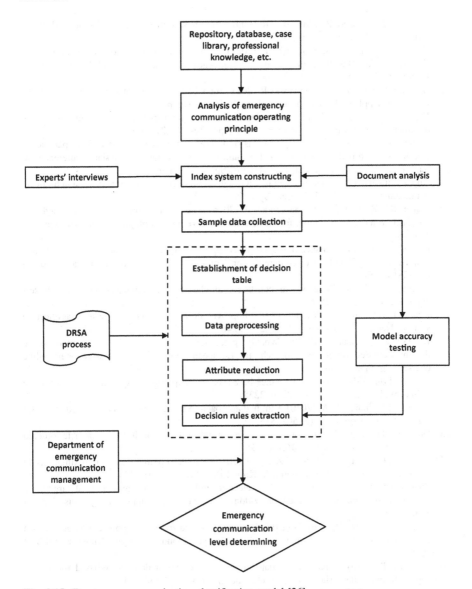

Fig. 8.15 Emergency communication classification model [26]

References

1. Słowiński R, Greco S, Matarazzo B (2007) Dominance-based rough set approach to multiple criteria decision support. Multiple criteria decision making/University of Economics in Katowice 2:9–56
2. Greco S, Matarazzo B, Słowiński R (1999) Rough approximation of a preference relation by dominance relations. Eur J Oper Res 117:63–83

3. Greco S, Matarazzo B, Słowiński R (2001) Rough sets theory for multicriteria decision analysis. Eur J Oper Res 129:1–47
4. Greco S, Matarazzo B, Slowinski R (2002) Multicriteria classification. In: Kloesgen W, Zytkow J (eds) Handbook of data mining and knowledge discovery, vol 1, no 9. Oxford University Press, pp 318–328 (chapter 16)
5. Błaszczynski J, Greco S, Słowinski R (2011) Inductive discovery of laws using monotonic rules, Eng Appl Artif Intell. https://doi.org/10.1016/j.engappai.2011.09.003
6. Hu Q, et al (2017) Spare parts classification in industrial manufacturing using the dominance-based rough set approach. Eur J Oper Res 262(3): 1136–1163
7. Mohamad M, Selamat A (2018) Analysis on hybrid dominance-based rough set parameterization using private financial initiative unitary charges data. In: Asian conference on intelligent information and database systems. Springer, Cham
8. Augeri MG et al (2010) Dominance-based rough set approach to budget allocation in highway maintenance activities. J Infrastruct Syst 17(2):75–85
9. Marin JC, Zaras K, Boudreau-Trudel B (2014) Use of the dominance-based rough set approach as a decision aid tool for the selection of development projects in Northern Quebec. Mod Econ 5(07):723
10. Pancerz K (2012) Dominance-based rough set approach for decision systems over ontological graphs. In: 2012 federated conference on computer science and information systems (FedCSIS). IEEE
11. Susmaga R (2014) Reducts and constructs in classic and dominance-based rough sets approach. Inf Sci 271:45–64
12. Azar AT, Inbarani HH, Devi KR (2017) Improved dominance rough set-based classification system. Neural Comput Appl 28(8):2231–2246
13. Bouzayane S, Saad I (2017) Weekly predicting the at-risk MOOC learners using dominance-based rough set approach. In: European conference on massive open online courses. Springer, Cham
14. Liou JJ, Tzeng GH (2010) A dominance-based rough set approach to customer behavior in the airline market. Inf Sci 180(11):2230–2238
15. Kusunoki Y, Inuiguchi M (2010) A unified approach to reducts in dominance-based rough set approach. Soft Comput 14(5):507–515
16. Du WS, Hu BQ (2016) Dominance-based rough set approach to incomplete ordered information systems. Inf Sci 346:106–129
17. Li S, Li T, Zhang Z, Chen H, Zhang J (2015) Parallel computing of approximations in dominance-based rough sets approach. Knowl Based Syst 87:102–111
18. Zhang HY, Yang SY (2017) Feature selection and approximate reasoning of large-scale set-valued decision tables based on α-dominance-based quantitative rough sets. Inf Sci 378:328–347
19. Augeri MG, Cozzo P, Greco S (2015) Dominance-based rough set approach: an application case study for setting speed limits for vehicles in speed controlled zones. Knowl Based Syst 89:288–300
20. Li S, Li T (2015) Incremental update of approximations in dominance-based rough sets approach under the variation of attribute values. Inf Sci 294:348–361
21. Li S, Li T, Liu D (2013) Incremental updating approximations in dominance-based rough sets approach under the variation of the attribute set. Knowl Based Syst 40:17–26
22. Li Y, Jin Y, Sun X (2018) Incremental method of updating approximations in DRSA under variations of multiple objects. Int J Mach Learn Cybernet 9(2):295–308
23. Luo C, et al (2015) Fast algorithms for computing rough approximations in set-valued decision systems while updating criteria values. Inf Sci 299:221–242

24. Shen KY, Tzeng GH (2015) A decision rule-based soft computing model for supporting financial performance improvement of the banking industry. Soft Comput 19(4):859–874
25. Maciag T, et al (2007) Evaluation of a dominance-based rough set approach to interface design. In: 2007 Frontiers in the convergence of bioscience and information technologies. IEEE
26. Zifu F, Hong S, Lihua W (2015) Research of the classification model based on dominance rough set approach for China emergency communication. Math Prob Eng (2015)

Chapter 9
Fuzzy Rough Sets

In this chapter, we will discuss the Fuzzy Rough Set Theory. Some core preliminaries will be presented along with necessary details. We will also discuss some state of the art fuzzy rough set approaches from the literature.

9.1 Fuzzy Rough Set Model

A particular use of Rough Set Theory is attribute reduction. This works when attributes have discrete values. However, in practical, attributes may have real values, which may not be precisely dealt with Rough Set Theory. Although Rough Set Theory provides a discretization process it results in loss of information. So this is the place where fuzzy rough set theory steps in.

A fuzzy rough set is the generalization of a rough set. In rough set, members may or may not belong to lower approximation. A member belonging to lower approximation with a membership of '1' is said to belong to approximated space with absolute certainty. However, this may not be the case of the fuzzy rough set where members may belong to lower and upper approximation with membership between 0 and 1. This provides greater help in dealing with uncertainty.

9.1.1 Fuzzy Approximations

A rough set can be expressed by a fuzzy membership function $\mu \to \{0, 0.5, 1\}$ to represent the negative, boundary and positive regions.

Here we provide formal definitions of fuzzy lower and upper approximations

© Springer Nature Singapore Pte Ltd. 2019
M. S. Raza and U. Qamar, *Understanding and Using Rough Set Based Feature Selection: Concepts, Techniques and Applications*, https://doi.org/10.1007/978-981-32-9166-9_9

$$\mu_{\underline{P}X}(F_i) = \inf \max\{1 - \mu_{F_i}(x), \mu_X(x)\} \forall I$$

$$\mu_{\bar{P}X}(F_i) = \inf \max\{\mu_{F_i}(x), \mu_X(x)\} \forall I$$

Here:

F_i= *equivalence class*
X = *fuzzy concept that requires approximation*

Note that although the universe of discourse in attribute selection is finite, this is not the case in general, hence the use of *sup* and *inf* [1]. Note that these definitions are a bit different from the lower and upper approximations (in crisp case) because the memberships of objects to both approximations is not clearly defined. As a result of this, the fuzzy lower and upper approximations are in redefined as [1]:

$$\mu_{\underline{p}x}(x) = \sup_{F \in U/P} \min(\mu_F(x), \inf_{y \in U} \max\{1 - \mu_{Fi}(x), \mu_X(y)\})$$

$$\mu_{\bar{p}x}(F_i) = \sup_{F \in U/P} \min(\mu_{Fi}(x), \sup_{y \in U} \min\{\mu_F(x), \mu_X(y)\})$$

As discussed earlier that fuzzy rough set is defined in terms of fuzzy lower and upper approximations, so the pair $\langle \underline{P}X, \bar{P}X \rangle$ is called fuzzy rough set.

9.1.2 Fuzzy Positive Region

In conventional rough set theory, the positive region is the union of the objects belonging to lower approximations of the decision classes. In fuzzy rough set theory, the positive region can be defined as

$$\mu_{POS_{P(Q)}}(x) = \sup_{x \in U/Q} \mu_{\underline{P}X}(x)$$

Object x will not belong to the positive region only if the equivalence class it belongs to is not a constituent of the positive region [1].

This also leads to the definition of fuzzy dependency function as follows:

$$\gamma_P(Q) = \frac{\mu_{POS_{P(Q)}}(x)}{|U|}$$

9.2 Fuzzy Rough Set Based Approaches

In [2], Qian et al. have presented an accelerator, called forward approximation, which combines sample reduction and dimensionality reduction together. This strategy is then used to optimize the heuristic process of fuzzy-rough feature selection. An improved feature selection algorithm is also designed on the basis of the accelerator. Using this accelerator, three representative heuristic fuzzy-rough feature selection algorithms are also enhanced. The proposed algorithm using forward feature selection starts with an empty set and keeps on adding attributes with maximal significance. The process continues until we get the Reduct.

Figure 9.1 shows the pseudocode of the fuzzy rough set based feature selection algorithm using forward approximation taken from [2].

In [3], authors have presented a fuzzy rough set based data reduction algorithm called 'Enhancing evolutionary instance selection algorithms by means of fuzzy rough set based feature selection (EIS-RFS)'. The algorithm results both in horizontal and vertical data reduction. Instance selection is performed using a steady-state genetic algorithm. It is then combined with a fuzzy rough set based feature selection process, which searches for the most interesting features to enhance both the evolutionary search process and the final preprocessed dataset.

Figure 9.2 shows the flowchart of the proposed algorithm presenting its main steps, which include:

Initialization (Step 1): In this step, chromosomes are initialized and initial subsets of features are selected.

Feature selection procedure (Step 4): This procedure consists of using the RST-based FS filter method, using as input the current best chromosome of the population are found out.

Input: Decision table $S = (U, C \cup D)$;
Output: One feature subset red.
Step 1: red $\leftarrow \emptyset$, i $\leftarrow$ 1, $R_1 \leftarrow$ red, $P_1 \leftarrow \{R_1\}$ and $U_1 \leftarrow U$; //red is the pool to conserve the selected attributes
Step 2: While $EF(red, D) \neq EF(C, D)$ Do //This provides a stopping criterion.
{
Compute the positive region of forward approximation $POS_{P_i}^{U}(D)$
$U_{i+1} \leftarrow U - POS_{P_i}^{U}(D)$
$i \leftarrow i + 1$,
$B \leftarrow C - red$,
Select $a_0 \in B$ which satisfies $Sig(a_0, red, D, U_i) = \max\{Sig(a_k, red, D, U_i), a_k \in B\}$,
If $Sig(a_0, red, D, U_i) > 0$, then red $\leftarrow$ red $\cup \{a_0\}$,
$R_i \leftarrow R_i \cup \{a_0\}$,
$P_i \leftarrow \{R_1, R_2, ..., R_i\}$;
}
Step 3: Return red and end.

Fig. 9.1 An improved feature selection algorithm based on the forward approximation (FA)

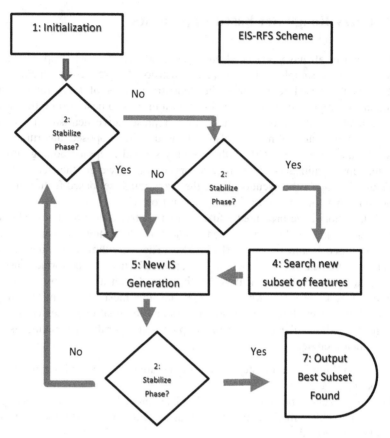

Fig. 9.2 Flowchart depicting the main steps of EIS-RFS. Rectangles depict processes whereas rhombuses depict decisions taken by the algorithm taken from [3]

Instance Selection procedure (Step 5): This process is carried out using the steady-state GA.

Output (Step 7): When the fixed number of evaluations runs out, the best chromosome of the population is selected as the best subset of instances found.

The rest of the operations (Steps 2, 3 and 6) control whether each of the former procedures should be carried out.

In [4], Jensen et al. have given a new fuzzy-rough based QuickReduct algorithm. The algorithm uses a new dependency function γ' to select the attribute that needs to be added to current Reduct set. The algorithm works in the same way as the original QuickReduct algorithm. It should be noted that in conventional rough set theory the Reduct set is the one having attributes with dependency equal to that of entire feature set and generally equal to '1' in case of consistent dataset, however, this may not be the case for a fuzzy rough approach as the uncertainty encountered when objects belong to many fuzzy equivalence classes results in a reduced total

FRQUICKREDUCT($\mathbb{C}$,$\mathbb{D}$).
$\mathbb{C}$, the set of all conditional features;
$\mathbb{D}$, the set of decision features.

(1) $R \leftarrow \{\}; \gamma'_{best} = 0; \gamma'_{prev} = 0$
(2)   **do**
(3) $T \leftarrow R$
(4) $\gamma'_{prev} = \gamma'_{best}$
(5) $\forall x \in (\mathbb{C} - R)$
(6)   **if** $\gamma'_{R \cup \{x\}}(\mathbb{D}) > \gamma'_T(\mathbb{D})$
(7) $T \leftarrow R \cup \{x\}$
(8) $\gamma'_{best} = \gamma'_T(\mathbb{D})$
(9) $R \leftarrow T$
(10)   **until** $\gamma'_{best} == \gamma'_{prev}$
(11)   **return** R

Fig. 9.3 The fuzzy-rough QuickReduct algorithm [4]

dependency. The alternate way may be to find the dependency of entire feature set and use it as the denominator to allow γ' to reach '1'. The fuzzy-rough QuickReduct algorithm is developed on these lines. The pseudocode of the algorithm is given in Fig. 9.3.

In [5], Wang et al. present a Fitting Model for Feature Selection with Fuzzy Rough Sets. Existing fuzzy-rough approaches use the fuzzy rough dependency to select features. However, this model can merely maintain a maximal dependency function. It does not fit a given data set well and cannot ideally describe the differences in sample classification. Therefore, the authors introduce a new model for handling this problem. First, they define the fuzzy decision of a sample using the concept of fuzzy neighborhood. Then, a parameterized fuzzy relation is introduced to characterize the fuzzy information granules, using which the fuzzy lower and upper approximations of a decision are reconstructed and a new fuzzy rough set model is introduced. This can guarantee that the membership degree of a sample to its own category reaches the maximal value. Furthermore, their approach can fit a given data set and effectively prevents samples from being misclassified. Finally, the authors define the significant measure of a candidate attribute and design a greedy forward algorithm for feature selection. Figure 9.4 shows the heuristic algorithm based on fitting the fuzzy rough set model.

In [6], Dai et al. present a feature selection based on information gain ratio in fuzzy rough set theory. The proposed algorithm was then used for tumor

Input: Decision table $< U, A, D >$, thresholds ε and λ // ε is
the threshold for the fuzzy neighborhood of a sample.
similarity. λ is the threshold for the fuzzy neighbor-
ood of decision D.

Output: One reduct *red*.

1: $\forall a \in A$: compute the relation matrix R_a;

2: Compute the fuzzy decision $\tilde{D} = \{\tilde{D}_1, \tilde{D}_2, \cdots \tilde{D}_r\}$;

3: Initialize: $red = \varnothing$, $B = A - red$, start = 1; // *red* is the pool
containing the selected attributes and B is for the
left attributes.

4: while start

5: $T \leftarrow \varnothing$

6:     for each $a_i \in B$

7:         $T \leftarrow red \cup \{a_i\}$;

8:         Compute fuzzy similarity relation R_T^ε.

9:         for each $x_j \in U$, suppose $x_j \in D_i$;

10:             Compute fuzzy lower approximation $\underline{R_T^\varepsilon(D_i)}(x_j)$.

11:         end for

12:         $\partial_{red \cup a_i}^\varepsilon (D) = \text{sum}(\ (\max_{D_i \in U/D} \underline{R_T^\varepsilon(D_i)})\ )/n$;

15:     end for

16:     Find attribute a_k with maximum value $\partial_{red \cup a_k}^\varepsilon (D)$.

17:     Compute $SIG^\varepsilon(a_k, red, D) = \partial_{red \cup a_k}^\varepsilon (D) - \partial_{red}^\varepsilon (D)$.

18:     if $SIG^\varepsilon(a_k, red, D) > 0$

19:         $red \leftarrow red \cup a_k$;

20:         $B \leftarrow B - red$;

21:     else

22:         start=0;

23:     end if

24: end while

25: return *red*

Fig. 9.4 Heuristic algorithm based on fitting fuzzy rough sets (NFRS) [5]

classification. Algorithm uses mutual information gain ratio for selecting attributes.
Given a fuzzy decision system FDS = {U, C ∪ D, V, f}, where C is the condition
attribute set and D is the decision attribute. B ⊆ C, $\forall a \in C - B$, the mutual
information gain ratio of attribute a, Gain Ratio(a, B, D) can be defined as

$$Gain\_Ratio\widetilde{(a,B,D)} = \frac{\widetilde{Gain\_Ratio}(a,B,D)}{\widetilde{H}(\{a\})}$$

$$= \frac{\widetilde{I}(B \cup \{a\} : D) - I(B : D)}{\widetilde{H}(\{a\})}$$

$$If\ B = \emptyset\ then\ Gain\_Ratio\widetilde{(a,B,D)} = \frac{\widetilde{I}(\{a\} : D)}{\widetilde{H}(\{a\})}$$

The pseudocode of the proposed algorithm is given below in Fig. 9.5.

In [7], Chen et al. have provided a heuristic algorithm to find reducts. The authors discuss Gaussian kernels as fuzzy T-similarity relations to develop Gaussian kernel-based fuzzy rough sets and consider attribute reduction with Gaussian kernels. They introduce Gaussian kernel into fuzzy rough sets for computing fuzzy similarity relation and develop a novel method of attribute reduction with parameter based on the proposed model. The authors also discuss the structure of subsets of selected attributes with fuzzy discernibility matrix. The authors argue that in real applications, it is not necessary to find all the Reducts. It is enough to address the real problem by using one of the Reducts. Figure 9.6 shows its heuristics algorithm to find Reducts.

Step 1. Let B =∅;
Step 2. For every attribute a ∈ C − B, compute the significance of condition attribute a, $Gain\_Ratio\widetilde{(a,B,D)}$
Step 3. Select the attribute which maximize the
$Gain\_Ratio\widetilde{(a,B,D)}$, record it as a; and B←B∪{a};
Step 4. If GainRatio(a, B, D) > 0, then B←B∪{a}, goto Step 2, else goto Step 5;
Step 5. The set B is the selected attributes.

Fig. 9.5 Attribute selection based on the gain ratio [6]

Input: (U,C,D), Reduct ← { }
Step 1: Compute the similarity relation of the set of all condition attributes: R_G^n
Step 2: Compute $Pos_C(D) = U_{t=1}^s R_G^n D_t$
Step 3: Compute c_{ij}
Step 4: Compute $Core_D(C) = \cup\{Q_{ij} \subseteq C : Q_{ij} = \cap\{P : {}^\wedge P\ 2\ c_{ij}\},\ i,j = 1,2,...,m\}$; Delete those c_{ij} with nonempty overlap with $Core_D(C)$;
Step 5: Let Reduct = $Core_D(C)$;
Step 6: Add the element a whose frequency of occurrence is maximum in all c_{ij} into Reduct; and delete those c_{ij} with
nonempty overlap with Reduct;
Step 7: If there still exist some cij ≠ ∅, go to Step 6; Otherwise, go to Step 8;
Step 8: If Reduct is not independent, delete the redundant elements in Reduct;
Step 9: Output Reduct.

Fig. 9.6 Heuristics algorithm to find Reducts [7]

The computational complexity of this algorithm is $O(|U|^2 * C)$.

In [8], Hu et al. have presented an information measure for computing discernibility power of a crisp equivalence relation or a fuzzy one, which is the key concept in the classical rough set model and fuzzy rough set model. Based on the information measure, a general definition of significance of nominal, numeric and fuzzy attributes was also presented. The authors redefine the independence of hybrid attribute subset, Reduct and relative Reduct. Then two greedy reduction algorithms for unsupervised and supervised data dimensionality reduction based on the proposed information measure are constructed. Figures 9.7 and 9.8 show the presented algorithms.

In [9], authors have introduced a simple hybrid attribute reduction algorithm based on a generalized fuzzy rough model. A theoretic framework of fuzzy rough model based on fuzzy relations is presented, which underlies a foundation for algorithm construction. The authors derive several attribute significance measures based on the proposed fuzzy rough model and construct a forward greedy algorithm for hybrid attribute reduction. Figure 9.9 shows the pseudo code of the algorithm:

Input: Information system IS $< U, A, V, f >$
Output: One reduct of IS
Step 1: $\forall a \in A$: compute the equivalence relation;
Step 2: $\emptyset \rightarrow$red;
Step 3: For each $a_i \in A - red$
Compute $H_i = H(a_i, red)$
End
Step 4: Choose attribute which satisfies:
$H(a|red) = \max_i(SIG(a_i, red))$
Step 5: If H(a_j|red) > 0, then red U a $\rightarrow$ red goto step 3
Else return red
End

Fig. 9.7 Algorithm for calculating Reduct [8]

Input: Information system IS $< U, A = C \cup d, V, f >$
Output: One relative reduct D_red of IS
Step 1: $\forall a \in A$: compute the equivalence;
Step 2: $\emptyset \rightarrow D\_red$
Step 3: For each $a_i \in A - red$
Compute $H_i = SIG(a_i, D\_red, d)$
End
Step 4: Choose attribute which satisfies:
$SIG(a, red, d) = \max_i(H_i)$
Step 5: If SIG(a,red,d) > 0, then D_red U $a \rightarrow$ D_red goto step 3
Else return, D_red
End

Fig. 9.8 Algorithm for calculating relative Reduct [8]

Input: Hybrid decision table $\langle U, A^c \cup A^r \cup d, V^c \cup V^r, f \rangle$ and Threshold k //A^c and A^r are categorical and numerical attributes

//k is the threshold for computing the lower approximations
Output: One reduct *red*.

Step 1: $\forall a \in A$:compute the equivalence relation R_a;

Step 2: $\phi \rightarrow red$; // *red* is the pool to contain the selected attributes

Step 3: For each $a_i \in A - red$
Compute $SIG(a_i, B, D) = \gamma_{red \cup a}^{kl}(D) - \gamma_{red}^{kl}(D)$, // Here we define $\gamma_{\emptyset}^{kl}(D) = 0$
end

Step 4: Select the attribute a_k which satisfies:

$$SIG(a_k, B, D) = \max_i(SIG(a_i, red, B))$$

Step 5: If $SIG(a_k, B, D) > 0$,

$red \cup a_k \rightarrow red$

go to step 3
else
return *red*
Step 6: end

Fig. 9.9 Forward attribute reduction based on variable precision fuzzy rough model (FAR-VPFRS) [9]

9.3 Summary

In this chapter, we have presented some basic concepts of fuzzy rough set theory and explained the difference between conventional rough sets and fuzzy rough sets. Fuzzy rough approximations and positive regions were presented. Then we discussed various fuzzy rough set based algorithms from different domains along with pseudocode of each algorithm.

References

1. Salama AS, Elabarby OG (2012) Fuzzy rough set and fuzzy ID3decision approaches to knowledge discovery in datasets. ISPACS
2. Qian Y et al (2015) Fuzzy-rough feature selection accelerator. Fuzzy Sets Syst 258:61–78
3. Derrac J et al (2012) Enhancing evolutionary instance selection algorithms by means of fuzzy rough set based feature selection. Inf Sci 186(1):73–92
4. Jensen R, Shen Q (2004) Semantics-preserving dimensionality reduction: rough and fuzzy-rough-based approaches. IEEE Trans Knowl Data Eng 16(12):1457–1471
5. Wang C et al (2017) A fitting model for feature selection with fuzzy rough sets. IEEE Trans Fuzzy Syst 25(4):741–753

6. Dai J, Qing X (2013) Attribute selection based on information gain ratio in fuzzy rough set theory with application to tumor classification. Appl Soft Comput 13(1):211–221
7. Chen D, Qinghua H, Yang Y (2011) Parameterized attribute reduction with Gaussian kernel based fuzzy rough sets. Inf Sci 181(23):5169–5179
8. Hu Q, Daren Yu, Xie Z (2006) Information-preserving hybrid data reduction based on fuzzy-rough techniques. Pattern Recogn Lett 27(5):414–423
9. Hu Q, Xie Z, Daren Yu (2007) Hybrid attribute reduction based on a novel fuzzy-rough model and information granulation. Pattern Recogn 40(12):3509–3521

Chapter 10
Introduction to Classical Rough Set Based APIs Library

In this chapter, we will provide some implementation of some basic functions of Rough Set Theory. Implementation of RST functions can be found in other libraries as well. The major aspect here is that source code is also provided with each and every line explained. The explanation in this way will help research community to not only easily use the code but also they can modify as per their own research requirements. We have used Microsoft Excel VBA to implement the function. The reason behind is that VBA provides easy implementation and almost any of the dataset can easily be loaded into the Excel. We will not only provide the implementation of some of the basic RST concepts but also complete implementation and explanation of the source code of some of the most common algorithms like PSO, GA, QUICK Reduct, etc.

10.1 A Simple Tutorial

Before going into the details of source code, we will first explain some basic statements of Excel VBA that are most commonly used. Here we will provide a very basic introduction about the syntax. For more details, we will recommend you to take some good tutorial.

10.1.1 Variable Declaration

Variables are declared with 'Dim' statement. For example, to declare a variable by the name 'Count' of type integer, we will use 'Dim' statement as follows:

Dim count as Integer

© Springer Nature Singapore Pte Ltd. 2019
M. S. Raza and U. Qamar, *Understanding and Using Rough Set Based Feature Selection: Concepts, Techniques and Applications*,
https://doi.org/10.1007/978-981-32-9166-9_10

Here, 'Dim' is keyword to declare variable, 'Count' is variable name and 'Integer' is data type of 'Count'.

10.1.2 Array Declaration

Just like variable declaration, arrays are declared with 'Dim' Keyword. For example, to declare a one-dimensional array by the name 'List', we will use 'Dim' statement as follows:

Dim List (2) As Integer

Here 'Matrix' is the name of array, '2' is the upper bound of array, i.e. the index of the last element. It should be noted that in Excel VBA arrays are zero-indexed, i.e. the first element has index '0' so above-defined array will have three elements. The same is the case with two-dimensional array. Arrays in Excel VBA are dynamic, i.e. we can change dimensions at runtime, however, for that you have to define empty array (i.e. array without specifying array size) as follows:

Dim List() As Integer
ReDim List(3)
List(0) = 2

However, note that every time you redefine array, previous data will be lost.

For two-dimensional array, the same syntax will be followed but specifying two indexes as follows:

Dim Matrix(3, 4) As Integer
Matrix(2, 2) = 3

Above two lines define a two-dimensional array and initialized the element in the third row and third column (remember that arrays are zero-indexed).

10.1.3 Comments

Comments are an important part of any programming language. In Excel, VBA comments are started by comma (') symbol. For example, following line will be commented:

'This is a comment and comments are turned green by default.

10.1.4   If–Else Statement

If–Else statement is a conditional statement used to implement branching. Its syntax is as follows:

If Count = 0 Then
'Statements here
Else
'Statements here
End If

After the keyword 'If', there is expression to be evaluated, 'Then' is the keyword used'. If the conditional expression solves to 'True', statements till 'Else' keyword will be executed otherwise statements after 'Else' will be executed. Finally 'End If' marks the end of 'If' condition.

10.1.5   Loops

The most common loops used are 'for-loop' and 'while-loop'. Below is the syntax of 'for-loop':

For Index = 1 To 10
count = count + 1
Next

Here 'For' is the keyword, 'Index' is the counter variable that starts with value '1', the loop will keep on iterating until the value of 'Index' remain less than or equal to '10'. Greater than '10' will cause the loop to terminate. 'Next' marks the end of the loop body.

For-loop works as counter, while-loop, on the other hand, keeps on iterating until a specific condition remains 'True'. Following is the syntax of while-loop:

While (i < 10)
i = i + 1
Wend

'While' is the keyword after which we have the expression to be evaluated. Set of statements till the keyword 'Wend' is called the body and is executed until conditional expression remains 'True'.

10.1.6 Functions

Functions are reusable code just like any other programming language. In Excel VBA, function definition has the following syntax:

```
Function Sum(ByVal x As Integer, ByVal y As Integer) As Integer
Dim Answer As Integer
Answer = x + y
Sum = Answer
End Function
```

'Function' is the keyword, then there is name of function, parameter list is specified in parentheses. Here 'ByVal' means 'By Value', to receive a parameter 'By Reference', we will use the keyword 'ByRef' while will let us modify the original variable in case if any change is made in the function. After parentheses, we specify the return type of function. In the above example, 'Sum' function will return an integer. Then we have the body of the function. To return a value from the function, it is assigned to the name of the function, e.g. here Sum = Answer' means that the value of the variable 'Answer' will be returned by function 'Sum'.

10.1.7 LBound and UBound Functions

LBound and UBound functions return the lower bound and upper bound of array. For example, for the array:

```
Dim List (3) as Integer
LBound (List) will return '0' and UBound (List) will return '3'.
```

10.2 How to Import the Source Code

You are given the '.bas' files that contain MS Excel VBA code. In order to use and modify the code, '.bas' files need to be imported in excel file. Here we will explain how to import '.bas' files in Excel.

To use the source code, you need any of the MS Excel 2013 or later version. However, note that the document should enable macros in order to run the code. To enable a macro, perform the following steps:

(1) Click File > Options (Fig. 10.1).
(2) Click 'Trust Center' from 'Excel Options' dialogue (Fig. 10.2).
(3) Click 'Trust Center Setting' button on the same dialog box (Fig. 10.3).

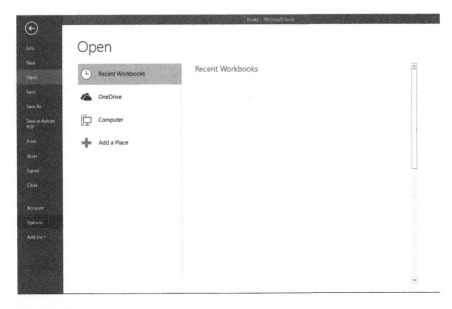

Fig. 10.1 File menu

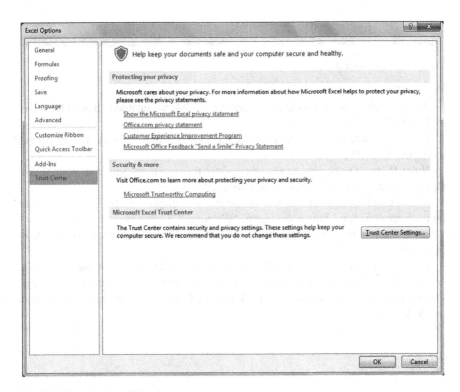

Fig. 10.2 Excel option dialog box

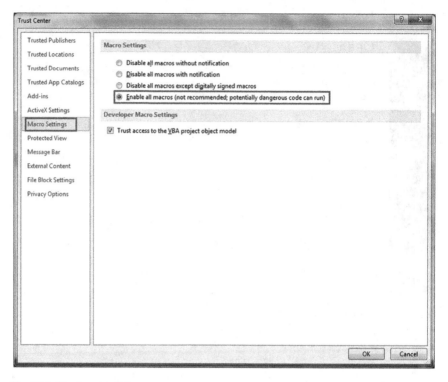

Fig. 10.3 Trust center dialogue

Alternatively, you can save the document as 'Macro enabled' from file 'Save' dialogue (Fig. 10.4).

To write/update source code, click the Visual Basic button on the Developer tab in ribbon (Fig. 10.5).

Note: If the 'Developer' tab is not visible then you can select it from 'Excel Options' dialog box from 'Customize Ribbon' tab (Fig. 10.6).

Now we will write a simple VBA code that will display 'Hello World' message on clicking command button captioned 'Welcome'.

Open Excel file and click on 'Developer' tab > Insert > Button as shown in Fig. 10.7.

Draw command button anywhere on the sheet. Once the button is drawn, 'Assign Macro' window will appear. For the time being, just cancel this window. Default caption of the button will be 'Button 1'. Click on it and change its caption to 'Welcome'.

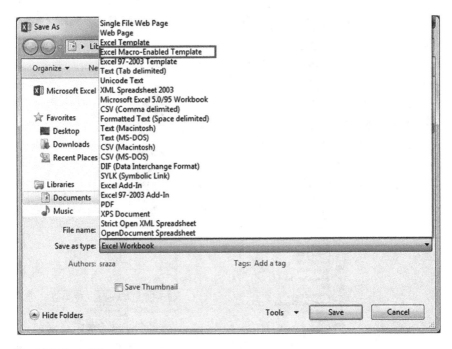

Fig. 10.4 Save dialogue

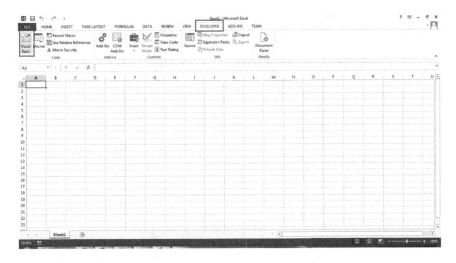

Fig. 10.5 Visual Basic button in Developer tab

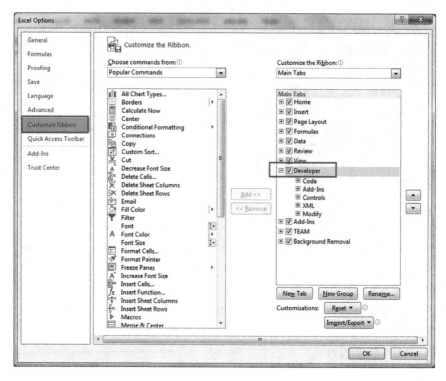

Fig. 10.6 Excel options dialogue

Fig. 10.7 Insert command button

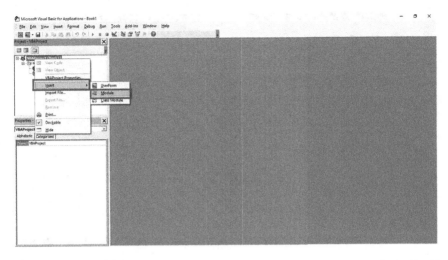

Fig. 10.8 Insert module

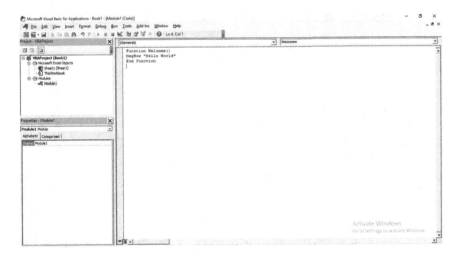

Fig. 10.9 Welcome function

Open visual basic editor, insert a module by right-clicking on 'Project', then 'Insert' and then 'Module' as shown in Fig. 10.8.

Once the module is inserted write down the following code as shown in Fig. 10.9.

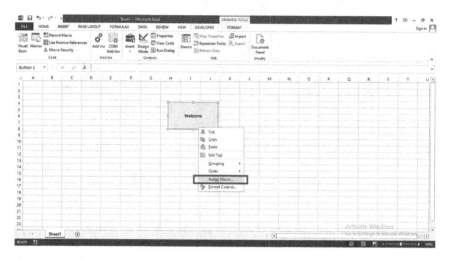

Fig. 10.10 Assign macro from popup menu

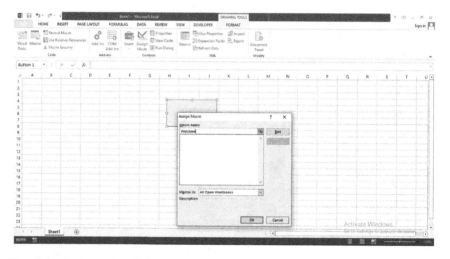

Fig. 10.11 Assign macro window

Now in Excel sheet right-click on the button and select 'Assign Macro' from the popup menu, assign 'Welcome' function to the button as shown in Figs. 10.10 and 10.11.

Click on the 'Ok' button. Code completed. Just click 'Welcome' button and message box will appear with 'Hello World1' message as shown in Fig. 10.12.

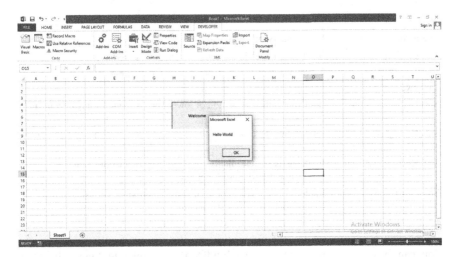

Fig. 10.12 Welcome message

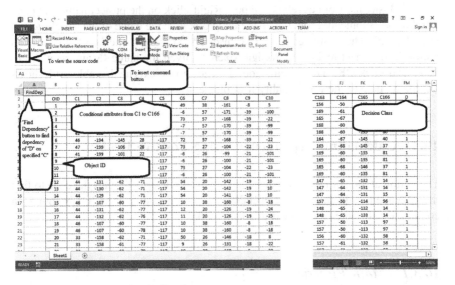

Fig. 10.13 How to store data in Excel

Here is how you will store dataset in excel sheet. We will explain the structure of file using 'Musk' dataset from (Fig. 10.13).

Actual dataset starts from 'C3' cell. Column 'B' specifies ObjectID of each row. Note that we have given integer numbers to each row to make code simple. The last column, i.e. 'FM' contains decision class (in case of Musk dataset).

Table 10.1 Source code files and tasks they implement

File name	Functionality implemented
Musk_I_Dep.bas	Calculates dependency using incremental classes
Musk_P_Dep.bas	Calculates dependency using the positive region
QuickReduct.bas	Implements QuickReduct algorithm
SpecHeart_I_LA.bas	Implements the calculation of lower approximation using dependency classes
SpecHeart_I_UA.bas	Implements the calculation of upper approximation using dependency classes
SpecHeart_P_LA.bas	Implements the calculation of upper approximation using conventional RST based method
SpecHeart_P_UA.bas	Implements the calculation of lower approximation using conventional RST based method

You can insert the command button ('FindDep' as shown in figure) from the 'Insert' tab on the developer ribbon. After the button is inserted.

Now all you need is to import the given source code. For this purpose open 'Visual Basic' tab from developer ribbon. Right-click on the project name and click the 'Import' submenu. From the open dialog box give the name of the '.bas' file and click 'Open'. The source code will be inserted. Now right-click on the command button in the sheet and click 'Assign Macro'. Given the name 'Main' as a macro name and click OK. Your source code is ready to be executed. Table 10.1 shows the file name and the functionality it implements.

10.3 Calculating Dependency Using Positive Region

The button at the left top of the sheet captioned 'FindDep' executes the code to find the dependency. In this case, Dependency of the decision class 'D' is found on conditional attributes mentioned in source code.

10.3.1 Main Function

Clicking the 'FindDep' button executes the 'Main' method. The code of the main method is given in the listing below (Listing 10.1).

Listing 10.1: Main Method
Row1:   Function Main()
Row2:   Dim i, j As Integer
Row3:   Rub = 6598
Row4:   Cub = 168
Row5:   ReDim data(1 To rub, 1 To cub)
Row6:   For i = 1 To rub
Row7:   For j = 1 To cub
Row8:   Data(i, j) = Cells(2 + i, 1 + j).Value
Row9:   Next j
Row10: Next i
Row11: ObjectID = LBound(data, 2)
Row12: DAtt = UBound(data, 2)
Row13: Dim Chrom() As Integer
Row14: ReDim Chrom(1 To 150)
Row15: For i = 2 To 10
Row16: Chrom(i - 1) = i
Row17: Next
Row18: ReDim TotalCClasses(1 To RUB, 1 To RUB + 1)
Row19: dp = CalculateDRR(chrom)
Row20: End Function

Now we will explain this code Row by Row. Row1 declares the main method. Then we have declared two variables i and j to use as an index of the loops. RUB and CUB stand for 'Row Upper Bound' and 'Column Upper Bound'. These two variables represent the maximum number of rows and columns in the dataset. Note that there are 6598 rows in musk and there are 166 conditional attributes, however, we will have to specify two extra attributes one for the decision class and one for the ObjectID. This will let us easily store data in our local array and refer to ObjectID just by specifying the first column of the array. Row5 declared an array called 'Data' equal to total dataset size including decision class and ObjectID. As interaction with Excel sheet is computationally too expensive at runtime, so we will load the entire dataset in the local array. Redim statement at Row5 specifies the size of the array. The array is initialized with the dataset from Row 6 to Row10.

Row11 and Row12, respectively, initialize ObjectID and DAtt (Decision Attribute). ObjectID is initialized with the index of the first column of 'Data' array while DAtt is initialized with the last index. Row14 defines an array named 'Chrom' which actually contains the indexes of attributes on which dependency of decision class 'D' is determined. Row15 to Row17 initializes chrom array with the attributes on whom dependency of 'D' is to be calculated. Note that Chrom will logically start from column 'C' in sheet, i.e. the first conditional attribute 'C1'. So far this means that you have assigned consecutive attributes to find dependency, however, you can assign selective attributes but in that case, you will have to initialize array manually without loop.

Row18 defines an array called 'TotalCClasses' which basically stores the equivalence class structure using the conditional attributes. Note that 'TotalCClasses' is defined globally and here it is assigned space dynamically. Finally, Row19 calls the function 'CalculateDRR(Chrom) calculates the dependency. Now we will explain 'CalculateDRR' method.

10.3.2   *CalculateDRR Function*

CalculateDRR function actually calculates dependency of decision class on conditional attributes. Listing 10.2 shows source code of this function, first, we will explain the working of this function and then all the functions called by it in sequence.

Listing 10.2: CalculateDRR Method
Row1:   Function CalculateDRR(ByRef chrom() As Integer) As Double
Row2:   Call SetDClasses
Row3:   Call ClrTCC
Row4:   Dim R As Integer
Row5:   Dim nr As Integer
Row6:   For R = 1 To RUB
Row7:   If (AlreadyExists(TotalCClasses, TCCRCounter, R) <> True) Then
Row8:   InsertObject Data(R, ObjectID), TotalCClasses, TCCRCounter
Row9:   For nr = R + 1 To RUB
Row10:   If (MatchCClasses(R, nr, chrom, TCCRCounter) = True) Then
Row11:   InsertObject Data(nr, ObjectID), TotalCClasses, TCCRCounter
Row12:   End If
Row13:   Next
Row14:   TCCRCounter = TCCRCounter + 1
Row15:   End If
Row16:   Next
Row17:   Dim dp As Integer
Row18:   Dim ccrx As Integer
Row19:   Dim ddrx As Integer
Row20:   dp = 0
Row21:   For ccrx = 1 To TCCRCounter
Row22:   For ddrx = 1 To TDCRCounter
Row23:   dp = dp + FindDep(TotalCClasses, TotalDClasses, ccrx, ddrx)
Row24:   Next
Row25:   Next
Row26:   CalculateDRR = dp / RUB
Row27:   End Function

Function accepts the reference of chrom array (or any array containing indexes of conditional attributes) and returns the dependency as a double value. The function first calls the 'SetDClasses' function which constructs equivalence class structure using decision attribute. Please refer to the description of SetDClasses Method in Sect. 10.3.3. Then it calls the function 'ClrTCC' which initializes the array used for calculating the equivalence class structure developed by using conditional attributes. Please refer to the description of ClrTCC Method in Sect. 10.3.5.

Then two local variables are declared with the name 'R' and 'NR' used as an index for current now and next row. From Row7 to Row16 function constructs equivalence class structure using conditional attributes. The first loop considers each record and checks whether it already exists in equivalence class structure or not. In case it does not exist (i.e. this object is the one with whom no other object has same attribute values), it is inserted in equivalence class structure. Second loop starts with the next object from the current object (selected in the first loop) and matches values of conditional attributes for all remaining objects. All the objects that match are inserted in the equivalence class structure. Please refer to the definitions of functions AlreadyExsits, InsertObject and MatchCClasses in Sects. 10.3.6, 10.3.7 and 10.3.8 respectively.

Row21 to Row25 calculate cardinality of the positive region, i.e. third step of dependency calculation. First loop iterates through each equivalence class in 'TotalCClasses' and sends it to function 'PosReg' along with each equivalence class stored in 'TotalDClasses'. PosReg function checks whether the equivalence class in 'TotalCClasses' is subset of 'TotalDClasses' or not. Once both loops complete, we have the cardinality of objects stored in variable 'dp'. Please refer to the definition of PosReg method in Sect. 10.3.9. The cardinality is then divided by total number of records in dataset to calculate and return actual dependency.

10.3.3 SetDClasses Method

This function calculates the equivalence class structure using decision attribute. Note that in calculating dependency using conventional positive region based dependency measure, this is the first step. Listing 10.3 shows source code of this function.

Listing 10.3: SetDClasses Method

```
Row1:    Function SetDClasses() As Integer
Row2:    Dim C As Integer
Row3:    Dim Row As Integer
Row4:    Dim Ind As Integer
Row5:    Const MaxR As Long = 2
Row6:    Dim X As Integer
Row7:    TDCRCounter = 2
Row8:    X = 5581
Row9:    ReDim TotalDClasses(1 To TDCRCounter, 1 To (X + 3))
Row10:   TotalDClasses(1, 1) = 0
Row11:   TotalDClasses(1, 2) = 5581
Row12:   TotalDClasses(1, 3) = 3
Row13:   TotalDClasses(2, 1) = 1
Row14:   TotalDClasses(2, 2) = 1017
Row15:   TotalDClasses(2, 3) = 3
Row16:       For Row = 1 To RUB ' construct decision classes
Row17:           Ind = FindIndex(Row, TotalDClasses, maxr)
Row18:           TotalDClasses(Ind, TotalDClasses(Ind, 3) + 1) = Data(Row, ObjectID)
Row19:           TotalDClasses(Ind, 3) = TotalDClasses(Ind, 3) + 1
Row20:       Next
Row21:   SetDClasses = 1
Row22:   End Function
```

Decision class	Instance Count	Instance Count	Instances		
1	3	3	X1	X3	X4
2	1	3	X2		
3	1	3	X5		

Fig. 10.14 Equivalence class structure using decision attribute

Row1 to Row7 define the local variables used. 'TDCRCounter' represents the total number of decision classes. In Musk dataset, there are two decision classes, so we assigned it with value '2'. MaxR represents the total number of rows in 'TotalDClasses' array. This array contains rows equal to the number of decision classes, i.e. one row for each decision class. In our case, there are two decision classes, so there will be two rows. The structure of 'TotalDClasses' array is as follows (Fig. 10.14).

First column specifies the decision class, second column specifies total number of objects in this equivalence class, and third column stores index of last object. The rest of the columns store objects in that decision class. Here in this case objects X1, X2 and X4 belong to decision class '1'.

The number of rows in 'TotalDClasses' array is equal to total number of decision classes and number of columns equal to the maximum number of objects in any decision class plus '3'.

Next the loop runs to create the equivalence class structure. It first finds the row number of the TotalDClasses array where decision class (of the current record) is stored. It then places the current object in the column next to the last object in the same row. Finally updates the total number of objects for that decision class and index of the last object.

10.3.4   FindIndex Function

This function finds the row number of the decision class of the current object in 'TotalDClasses' array. Code of the function is given below in Listing 10.4.

Listing 10.4: FindIndex Method	
Row1:	Function FindIndex(R As Integer, ByRef TDC() As Integer, MaxR As Integer) As
Row2:	Integer
Row3:	Dim C As Integer
Row4:	For C = 1 To MaxR
Row5:	If Data(R, DAtt) = TDC(C, 1) Then
Row6:	Exit For
Row7:	End If
Row8:	Next
Row9:	FindIndex = C
Row10:	End Function

Function takes three Arguments, i.e. row number of current object, reference of 'TotalDClasses' array and maximum number of rows in it. It then navigates through the first column of each row (where decision class is stored). If the decision class is matched with decision class of current object (whose ID is represented by variable R) then this object is returned. Note that if decision class of current object is not matched with any in 'TotalDClasses' then index of last row is returned where this object will be stored.

10.3.5   ClrTCC Function

Listing 10.5 below shows the source code of ClrTCC method

Listing 10.5: ClrTCC Method	
Row1:	Function ClrTCC() As Integer
Row2:	Dim i As Integer
Row3:	TCCRCounter = 0
Row4:	For i = 1 To RUB
Row5:	TotalCClasses(i, 1) = 0
Row6:	Next
Row7:	ClrTCC = 1
Row8:	End Function

Row1 defines the function. Variable 'i' will be used as index variable in the loop. First, let's explain the structure of how 'TotalCClasses' stores equivalence class structure as shown in Fig. 10.15. Each row stores an equivalence class, e.g. objects X1 and X3 belong to the same equivalence class whereas objects X2, X4, and X5 represent another equivalence class. First column shows the total number of

Fig. 10.15 Equivalence class structure using conditional attributes

2	X1	X3		
3	X2	X4	X5	

objects in an equivalence class, e.g. in first row there are two objects in equivalence class whereas in second row there are three objects.

From Listing 7.1 note that the total number of rows in 'TotalCClasses' is equal to the total number of rows in the dataset. This is done for the worst (very rare) cases where each object has its own decision class. Similarly, the number of columns are equal to the number of rows plus one (1), this is again for worst case where all objects belong to the same class. The extra column stores the total number of objects in each equivalence class.

Now the function 'ClrTCC()' clears the first column of the 'TotalCClasses' array. It then returns the value '1' to indicate that function has been successfully executed.

10.3.6 AlreadyExists Method

This method checks if an object already exists in equivalence classes. Following is the listing of this function (Listing 10.6):

Listing 10.6: AlreadyExists Method
Row1:   Function AlreadyExists(ByRef TCC() As Integer, TCCRC As Integer, R As
Row2:   Integer) As Boolean
Row3:   Dim i As Integer
Row4:   Dim j As Integer
Row5:   Dim Exists As Boolean
Row6:   Exists = False
Row7:   If TCCRC = 0 Then
Row8:   Exists = False
Row9:   End If
Row10:   For i = 1 To TCCRC
Row11:   For j = 2 To TCC(i, 1) + 1
Row12:   If (TCC(i, j) = Data(R, ObjectID)) Then
Row13:   Exists = True
Row14:   Exit For
Row15:   End If
Row16:   Next
Row17:   If (Exists = True) Then
Row18:   Exit For
Row19:   End If
Row20:   Next
Row21:   AlreadyExists = Exists
Row22:   End Function

The function takes three arguments, the reference of 'TotalCClasses' array, total number of rows in this array and row number (stored in 'R') that is to be checked in equivalence classes. The function first checks if 'TotalCClasses' array is empty in which case function returns 'False'. Then function runs two nested loops to check the objects column wise in each row. Note that the second loop starts with j = 2 because in equivalence class array first column stores total number of objects in decision class and actual objects in an equivalence class start from the second column. If the object is found at any location function returns 'True' else 'False'.

10.3.7 InsertObject Method

This method inserts an object in Equivalence class structure, i.e. 'TotalCClasses'. This function is called when the object does not exist in the array, so it inserts the object in next row and initializes the first column in this row equal to one (1). Following is the listing of this function (Listing 10.7).

Listing 10.7: InsertObject Method	
Row1:	Function InsertObject(O As Integer, ByRef TCC() As Integer, TCCRC As Integer)
Row2	TCC(TCCRC + 1, 1) = TCC(TCCRC + 1, 1) + 1
Row3:	TCC(TCCRC + 1, TCC(TCCRC + 1, 1) + 1) = O
Row4:	End Function

Function takes three arguments, ObjectID of the object to be inserted, 'TotalCClasses' array and total number of rows in the array so far having objects.

10.3.8 MatchCClasses Function

This function matches two objects according to the value of their attributes. It takes three arguments. Following is the listing of this function (Listing 10.8).

Listing 10.8: MatchCClasses Method	
Row1:	Function MatchCClasses(R As Integer, NR As Integer, ByRef chrom() As Integer,
Row2	TCCRC As Integer) As Boolean
Row3:	Dim j As Integer
Row4:	Dim Ci As Integer
Row5:	Dim ChromSize As Integer
Row6:	Dim ChromMatched As Boolean
Row7:	ChromMatched = True
Row8:	For Ci = LBound(chrom) To UBound(chrom)
Row9:	If (Data(R, chrom(Ci)) <> Data(NR, chrom(Ci))) Then
Row10:	ChromMatched = False
Row11:	Exit For
Row12:	End If
Row13:	Next
Row14:	MatchCClasses = ChromMatched
Row15:	End Function

Function takes three arguments, ObjectIDs of the objects to be compared and reference of 'Chrom' array which contains attributes for which objects need to be compared. It runs the loop that matches the objects against the values of the attributes stored in Chrom array. If both objects have the same values against attributes mentioned in chrom then it returns 'True' else 'False'.

10.3.9   PosReg Function

This function calculates cardinality of the objects belonging to the positive region. The listing of function is given below (Listing 10.9):

Listing 10.9: PosReg Method
Row1:   Function PosReg(ByRef TCC() As Integer, ByRef TDC() As Integer, cr As
Row2:   Integer, dr As Integer) As Integer
Row3:   Dim X, cnt, dpc, y As Integer
Row4:   dpc = 0
Row5:   cnt = TCC(cr, 1)
Row6:   If (TCC(cr, 1) <= TDC(dr, 2)) Then
Row7:   For X = 2 To TCC(cr, 1) + 1
Row8:   For y = 4 To TDC(dr, 3)
Row9:   If (TCC(cr, X) = TDC(dr, y)) Then
Row10:   dpc = dpc + 1
Row11:   End If
Row12:   Next
Row13:   Next
Row14:   End If
Row15:   If cnt = dpc Then
Row16:   PosReg = dpc
Row17:   Else
Row18:   PosReg = 0
Row19:   End If
Row20:   End Function

This function takes four arguments, i.e. reference of 'TotalCClasses' array, 'TotalDClasses' array, current row in 'TotalClasses' and current row in 'TotalDClasses'. The function then calculates the cardinality of objects in equivalence class stored in 'TotalCClasses' (at row number 'cr') which are a subset of equivalence class sored in 'TotalDClasses' (at row number 'dr'). Note that first loop at Row7 starts with $X = 2$ because in 'TotalCClasses' objects in each equivalence class start from index 2. Similarly, in 'TotalDClasses' indexes of objects start from index 4.

10.4 Calculating Dependency Using Incremental Dependency Classes

Now, we will explain how dependency can be calculated using IDCs. The data will be stored in excel file same as we stored in case of positive region based dependency measure.

10.4.1 Main Function

Clicking the 'FindDep' button will call 'Main' Method. Loading the data in the local array 'Data' and other variables is same as discussed before. Following is the listing of the Main function (Listing 10.10)

```
Listing 10.10: Main Method
Row1:   Function Main()
Row2:   Dim i, j As Integer
Row3:   Rub = 6598
Row4:   Cub = 168
Row5:   ReDim data(1 To rub, 1 To cub)
Row6:   For i = 1 To rub
Row7:      For j = 1 To cub
Row8:   Data(i, j) = Cells(2 + i, 1 + j).Value
Row9:      Next j
Row10:  Next i
Row11:  ObjectID = LBound(data, 2)
Row12:  DAtt = UBound(data, 2)
Row13:  Dim Chrom() As Integer
Row14:  ReDim Chrom(1 To 150)
Row15:  For i = 2 To 10
Row16:  Chrom(i - 1) = i
Row17:  Next
Row18:  dp = CalculateDID(chrom)
Row19:  End Function
```

All of the function is same as discussed before, however, at Row18, the function CalculateDID will calculate dependency using IDCs.

10.4.2 CalculateDID Function

CalculateDID calculates dependency using dependency classes. Listing 10.11 shows the source code of the function.

Listing 10.11: CalculateDID Method

```
Row1:    Function calculateDID(ByRef chrom() As Integer) As Double
Row2:    Dim DF As Integer
Row3:    Dim UC As Integer
Row4:    Dim ChromSize As Integer
Row5:    Dim FoundInGrid As Boolean
Row6:    Dim i As Integer
Row7:    Dim GRC As Integer
Row8:    Dim ChromMatched As Boolean
Row9:    Dim DClassMatched As Boolean
Row10:   Dim ChromMatchedAt As Integer
Row11:   Dim DClassMatchedAt As Integer
Row12:   GridRCounter = 0
Row13:   FoundInGrid = False
Row14:   ChromMatched = False
Row15:   DClassMatched = False
Row16:   ChromSize = UBound(chrom) - LBound(chrom) + 1
Row17:   DECISIONCLASS = UBound(chrom) + 1
Row18:   INSTANCECOUNT = DECISIONCLASS + 1
Row19:   AStatus = INSTANCECOUNT + 1
Row20:   ReDim Grid(1 To ChromSize + 3, 1 To RUB)
Row21:   If (GridRCounter = 0) Then
Row22:      GridRCounter = Insert(GridRCounter, chrom, 1)
Row23:      DF = 1
Row24:      UC = 1
Row25:   End If
Row26:   For i = 2 To RUB
Row27:      ChromMatchedAt = MatchChrom(i, chrom, ChromMatched, GridRCounter)
Row28:      If (ChromMatched = True) Then
Row29:         If (Grid(AStatus, ChromMatchedAt) <> 1) Then
Row30:            DClassMatched = MatchDClass(i, ChromMatchedAt)
Row31:            If (DClassMatched = True) Then
Row32:               DF = DF + 1
Row33:               Grid(INSTANCECOUNT, ChromMatchedAt) =
                     Grid(INSTANCECOUNT, ChromMatchedAt) + 1
Row34:            Else
Row35:               DF = DF - Grid(INSTANCECOUNT, ChromMatchedAt)
Row36:               Grid(AStatus, ChromMatchedAt) = 1
Row37:            End If
Row38:         End If
Row39:      Else
Row40:         GridRCounter = Insert(GridRCounter, chrom, i)
Row45:         DF = DF + 1
Row46:      End If
Row47:      UC = UC + 1
Row48:   Next
Row49:   calculateDID = DF / UC
         End Function
```

Function takes reference of integer array 'Chrom' as input. As discussed earlier, 'Chrom' contains indexes of the conditional attributes on whom the dependency of 'D' is to be determined. 'CalculateDID' uses 'Grid' as an intermediate data structure to calculate dependency. 'Grid' is just a two-dimensional array with

Table 10.2 Sample decision system

U	State	Qualification	Job
x_1	S1	Doctorate	Yes
x_2	S1	Diploma	No
x_3	S2	Masters	No
x_4	S2	Masters	Yes
x_5	S3	Bachelors	No
x_6	S3	Bachelors	Yes
x_7	S3	Bachelors	No

number of rows equal to number of records in dataset and number of columns equal to total number of conditional attributes on whom dependency is to be determined plus '3', e.g. if dependency is to be determined on attributes 'A1, A3 and A3', the number of columns should be equal to '6'. The last column specifies the status of attribute set that either these values of attributes have already been considered or not. The second last column specifies the total number of occurrences of the current value set in the dataset (so that if the same value of these attributes leads to a different decision class later, we may subtract this number from current dependency value. Third last column specifies the decision class and the first n-3 columns specify the values under the current attribute set. This matrix is filled for all the instances in the dataset. So for the dataset given in Table 10.2.

Contents of the grid (Fig. 10.16) will be as follows (for first three objects).

In Row22, we insert the first record in the Grid. Please refer to the definition of 'Insert' method (Sect. 10.3.7) about how to insert a record. 'DF' and 'UC' store 'Dependency Factor' and 'Universe Count'. Dependency factor stores the total number of objects corresponding to a positive region in the conventional method. As so far we have only inserted the first record so both DF and UC will be equal to '1'. Next method starts a loop from $i = 2$ (because we have already inserted the first record).

In Row27, MatchChrom function matches that if any object with same attribute values as that of the object 'i' already exists in Grid or not. In case if object with the same value already exists in grid, we will check that whether we already have considered it or not (i.e. either we have updated DF on the basis of this object or not), if this object has not been considered previously, we will match the decision class of the object with the one stored in Grid. For this purpose 'MatchDClass' function is used. If the decision class is also matched, we increment the 'DF' factor which means that this object belongs to the positive region. In Row33, we increment the Instance count (second last) column. However, decision class does not match then we will decrement the DF by the factor equal to the total number of previous occurrences of all objects with the same attribute values. We will also set last column (AStatus) to '1' to indicate that object has already been considered.

Fig. 10.16 Grid for
calculating IDC

S1	Doctorate	Yes	1	1
S1	Diploma	No	1	1
S2	Masters	No	1	1

However, if this object has already been considered, we will simply increment the
UC factor. In case if no object with same attributes values already exist in Grid, we
will insert this object using 'Insert' function and will update 'DF'.

Finally in Row49 we calculate and return actual dependency value.

10.4.3   Insert Method

This method inserts an object in Grid. Listing 10.12 shows the source code of this
method.

```
Listing 10.12: Insert Method
Row1:     Function Insert(GRC As Integer, chrom() As Integer, drc As Integer) As Integer
Row2:     Dim Ci As Integer
Row3:     For Ci = 1 To UBound(chrom)
Row4:         Grid(Ci, GRC + 1) = Data(drc, chrom(Ci))
Row5:     Next
Row6:     Grid(Ci, GRC + 1) = Data(drc, DAtt)
Row7:     Grid(Ci + 1, GRC + 1) = 1 'instance count
Row8:     Grid(Ci + 2, GRC + 1) = 0 'status
Row9:         Insert = GRC + 1
Row10:    End Function
```

The function receives three arguments, i.e. Grid Record count (GRC) is the total
number of records already inserted in Grid (note that Insert method is called when a
record is inserted in the grid for the first time, so we need to insert it in the next row
after the last inserted row in the grid). It then iterates through all the attribute
indexes stored in Chrom array and stores them in first n-3 columns of the Grid. In
third last column, decision class is stored, in the second last column instance count
is set to '1' as the record with such attribute values appear in the Grid for the first
time and finally status is set to '0' means that this object has not been
considered before.

10.4.4   MatchChrom Method

MatchCrhom method compares the attribute values of the current object with all those in the Grid. Following is the listing of this method (Listing 10.13).

Listing 10.13: MatchChrom Method
Row1:   Function MatchChrom(i As Integer, ByRef chrom() As Integer, ByRef ChromMatched As Boolean, GRC As Integer) As Integer
Row2:   Dim j As Integer
Row3:   Dim Ci As Integer
Row4:   For j = 1 To GRC
Row5:   ChromMatched = True
Row6:   For Ci = 1 To UBound(chrom)
Row7:   If (Data(i, chrom(Ci)) <> Grid(Ci, j)) Then
Row8:   ChromMatched = False
Row9:   Exit For
Row10: End If
Row11:   Next
Row12:   If (ChromMatched = True) Then
Row13: Exit For
Row14:   End If
Row15:   Next
Row16: MatchChrom = j
Row17: End Function

It receives three arguments, i.e. Grid Record Count (GRC), reference of the chrom array and Data Record Count (drc) that stores the index of the object whose attribute values need to be compared. The function iterates through rows, starting from first row in Grid, it compares attribute values with that of current object, first the attribute values stored in the first row are compared, then that in the second row and so on. At any row number where attribute values match, that row number is returned.

10.4.5   MatchDClass Method

In CalculateDID method, MatchDClass is called if attribute values are compared and the row number where the attribute values are compared and returned. Now at that row in the grid, MatchDClass compares the decision class stored with that of the current object. Following is the listing of this function (Listing 10.14)

Table 10.3 Variables and their description

Variable	Datatype	Purpose
Data()	Integer	Integer array that stores data. First column of the array stores ObjectID and last column stores decision class
ObjectID	Integer	Refers to the index of the first column in 'Data' array. Used to refer to a particular object
DAtt	Integer	Stores index of column containing decision class
RUB	Integer	Stores count of the total number of rows in the dataset
CUB	Integer	Stores the total number of columns in dataset
TotalDClasses ()	Integer	Stores equivalence class structure using decision class
TotalCClasses ()	Integer	Stores equivalence class structure using conditional attributes
TDCRCounter	Integer	Total number of equivalence classes in 'TotalCClasses' array
TCCRCounter	Integer	Total number of equivalence classes in 'TotalDClasses' array

---

Listing 10.14: MatchDClass Method

```
Row1:    Function MatchDClass(i As Integer, ChromMatchedAt As Integer) As Boolean
              Dim DCMatched As Boolean
Row2:    DCMatched = False
Row3:       If (Data(i, DAtt) = Grid(DECISIONCLASS, ChromMatchedAt)) Then
Row4:          DCMatched = True
Row5:       End If
Row6:       MatchDClass = DCMatched
Row7:    End Function
```

It receives two arguments, i.e. row number of the object in the dataset and the row number in grid where the chrom was matched. If decision class matches with that of object in dataset, then the function returns 'TRUE' else 'FALSE'.

Table 10.3 shows the important variables and their description.

10.5   Lower Approximation Using Conventional Method

Following function can be used to calculate lower approximation using conventional indiscernibility based method.

10.5.1   Main Method

Listing 10.15 shows the source code of the Main method.

Listing 10.15: Main Method

```
Row1:      Function Main() As Integer
Row2       GridRCounter = 0
Row3:      Dim i, j As Integer
Row4:      RUB = 80
Row5:      CUB = 46
Row6:      ReDim data(1 To RUB, 1 To CUB)
Row7:      For i = 1 To RUB
Row8:         For j = 1 To CUB
Row9:            data(i, j) = Cells(2 + i, 1 + j).Value
Row10:        Next j
Row11:     Next i
Row12:     OID = LBound(data, 2)
Row13:     DAtt = UBound(data, 2)
Row14:     Dim chrom() As Integer
Row15:     ReDim chrom(1 To CUB - 2)
Row16:     For i = 2 To CUB - 1
Row17:     chrom(i - 1) = i
Row18:     Next
Row19:     CalculateLAObjects chrom, 1
Row20:     Dim str As String
Row21:     For i = 1 To UBound(LAObjects)
Row22:        If LAObjects(i) <> 0 Then
Row23:        str = str & ",X" & LAObjects(i) & ""
Row24:
Row25:        End If
Row26:     Next
Row27:     Cells(2, 1).Value = str
Row28:     Main = 1
Row29:     End Function
Row30:
```

Function loads the data in the same way as we have previously explained. At Row19 it calls 'CalculateLAObjects' to calculate objects belonging to lower approximation. The function fills 'LAObjects' array with all such objects. This array contains indexes (record number) of objects belonging to lower approximation. Row21 to Row26 iterates through each object and stores them in a string which is then shown in Cell 'A2'. This is just for the display purpose, however, for any further processing on objects belonging to lower approximation 'LAObjects' can be used.

10.5.2   CalculateLAObjects Method

This method calculates the equivalence class structure using conditional attributes. Following is the listing of this method (Listing 10.16).

Listing 10.16: CalculateLAObjects Method
Row1:   Function CalculateLAObjects(ByRef chrom() As Integer, ByRef xc as Integer)
Row2:   Dim i As Integer
Row3:   Dim j As Integer
Row4:   Dim found As Integer
Row5:   SetDConcept xc
Row6:   Dim TotalCClasses() As Integer
Row7:   Dim TCCRCounter As Integer
Row8:   TCCRCounter = 0
Row9:   ReDim TotalCClasses(1 To RUB, 1 To RUB + 1)
Row10:   Dim R As Integer
Row11:   Dim nr As Integer
Row12:   For R = 1 To RUB
Row13:   If (AlreadyExists(TotalCClasses, TCCRCounter, R, data) <> True) Then
Row14:   InsertObject data(R, OID), TotalCClasses, TCCRCounter
Row15:   For nr = R + 1 To RUB
Row16:   If (MatchCClasses(R, nr, chrom, TCCRCounter) = True) Then
Row17:   InsertObject data(nr, OID), TotalCClasses, TCCRCounter
Row18:   End If
Row19:   Next
Row20:   TCCRCounter = TCCRCounter + 1
Row21:   End If
Row22:   Next
Row23:   Dim ccrx As Integer
Row24:   ReDim LAObjects(1 To RUB + 1)
Row25:   LAObjects(1) = 0
Row26:   For ccrx = 1 To TCCRCounter
Row27:   FindLAO TotalCClasses, ccrx
Row28:   Next
Row29:   ccrx = 0
Row30:   End Function

It takes two arguments, the 'Chrom' array, which contains indexes of the attributes under consideration and the 'XConcept'. 'XConcept' specifies the concept (in simple words, the value of decision class) for which we need to determine lower approximation. First, the function calls 'SetDConcept' method and passes it the value of 'XConcept' (Please refer to the definition of 'SetDConcept' method in Sect. 10.5.4). It then creates an array 'TotalCClasses' in the same way as discussed in previous sections. From Row12 to Row22 it constructs the equivalence class structure using conditional attributes mentioned in 'Chrom' array. The variable 'TCCRcounter' stores the total number of equivalence classes in 'TotalCClasses' array. It then calls the function FinLAO to which actually performs the third step of finding lower approximations, i.e. all the objects in equivalence class structure (using conditional attributes) which are subset of equivalence class structure (using decision attribute) of XConcept. For example, if there are three equivalence classes in 'TotalCClasses' stored in three rows, then the 'FindLAO' will be called for each equivalence class. Following is the description of 'FindLAO' method.

10.5.3 FindLAO Method

Find 'LAO' method actually finds the objects belonging to lower approximation. Following is the listing of this method (Listing 10.17)

Listing 10.5: FindLAO Method
Row1:   Function FindLAO(ByRef TCC() As Integer, cr As Integer)
Row2   Dim X, cnt, lac, y As Integer
Row3:   lac = 0
Row4:   cnt = TCC(cr, 1)
Row5:   If (TCC(cr, 1) <= XDConcept(1)) Then
Row6:   For X = 2 To TCC(cr, 1) + 1
Row7:   For y = 2 To XDConcept(1) + 1
Row8:   If (TCC(cr, X) = XDConcept(y)) Then
Row9:   lac = lac + 1
Row10:   End If
Row11:   Next
Row12:   Next
Row13:   End If
Row14:   If cnt = lac Then
Row15:   For X = 2 To TCC(cr, 1) + 1
Row16:   LAObjects(1) = LAObjects(1) + 1
Row17:   LAObjects(LAObjects(1) + 1) = TCC(cr, X)
Row18:   Next
Row19:   End If
Row20:   End Function

It takes two arguments, i.e. reference of 'TotalCClasses' array and the row number, i.e. the row index of the equivalence class structure which needs to be determined either as a subset of 'XConcept' or not. In Row4, it first determines the total number of objects in the current equivalence class structure, if the number of objects in current equivalence class structure are less than those in equivalence class structure using decision attribute, then it runs two nested for loops, first loop picks and object and iterates through equivalence structure of 'XConcept', if at any location the object is found, we increment lower approximation count (lac) variable. Finally, we check if this count is less than the total number of objects belonging to XConcept, or not. In case if total number of objects in current equivalence class are less than those belonging to XConcept, it means that set of objects in this equivalence class is subset of XConcept and thus belongs to lower approximation. Then we copy all these objects to 'LAObjects' array.

10.5.4   SetDConcept Method

This method constructs the equivalence class structure using XConcept. Following is the listing of this function (Listing 10.18).

Listing 10.18: SetDConcept Method

Row1:	Function SetDConcept (ByRef xc as Integer) As Integer
Row2	Dim XConcept As Integer
Row3:	Dim i As Integer
Row4:	XConcept = xc
Row5:	ReDim XDConcept(1 To RUB + 1)
Row6:	Dim CDCI As Integer
Row7:	CDCI = 2
Row8:	XDConcept(1) = 0
Row9:	For i = 1 To RUB
Row10:	If (data(i, DAtt) = XConcept) Then
Row11:	XDConcept(1) = XDConcept(1) + 1
Row12:	XDConcept(CDCI) = data(i, OID)
Row13:	CDCI = CDCI + 1
Row14:	End If
Row15:	Next
Row16:	SetDConcept = 1
Row17:	End Function

Function takes single argument, i.e. the XConcept value for which we need to construct equivalence class structure. The function then declares an XDConcepts array which will store the objects belonging to XConcept. Note that it is a one-dimensional array with a size equal to the total number of records in dataset plus 1 (in case if all objects in a dataset belong to the same class). The extra column (first one) stores the total number of objects in XDConcepts. This function traverses through all the records and stores index of those object (in XDConcepts array) for whom decision class values match XConcept. It also updates the first index of SDConcepts array to represent total number of objects in it.

10.6   Lower Approximation Using Redefined Preliminaries

Lower approximation is calculated by 'CalculateLAI' method. This method calculates lower approximation using redefined preliminaries. Following is the listing of this function (Listing 10.19):

	Listing 10.19: CalculateLAI Method
Row1:	Function CalculateLAI(ByRef chrom() As Integer, ByVal xc As Integer) As Single
Row2:	Dim DF, UC As Integer
Row3:	Dim i As Integer
Row4:	Dim GRC As Integer
Row5:	Dim ChromMatched As Boolean
Row6:	Dim DClassMatched As Boolean
Row7:	Dim ChromMatchedAt As Integer
Row8:	Dim DClassMatchedAt As Integer
Row9:	Dim NObject As Integer
Row10:	ReDim Grid(1 To cub + 3, 1 To rub)
Row11:	GridRCounter = 0
Row12:	ChromMatched = False
Row13:	DClassMatched = False
Row14:	ChromSize = UBound(chrom) - LBound(chrom) + 1
Row15:	DECISIONCLASS = UBound(chrom) + 1
Row16:	INSTANCECOUNT = DECISIONCLASS + 1
Row17:	AStatus = INSTANCECOUNT + 1
Row18:	XConcept = xc
Row19:	TDO = 0
Row20:	For i = 1 To rub
Row21:	If (data(i, DAtt) = XConcept) Then
Row22:	TDO = TDO + 1
Row23:	End If
Row24:	Next
Row25:	ReDim Grid(1 To ChromSize + 3 + TDO, 1 To rub) ' column,row
Row26:	If (GridRCounter = 0) Then
Row27:	GridRCounter = Insert(GridRCounter, chrom, 1) ' 1 represents first record
Row28:	End If
Row29:	For i = 2 To rub
Row30:	ChromMatchedAt = MatchChrom(i, chrom, ChromMatched, GridRCounter)
Row31:	If (ChromMatched = True) Then
Row32:	Grid(INSTANCECOUNT, ChromMatchedAt) = Grid(INSTANCECOUNT,
Row33:	ChromMatchedAt) + 1
	If (Grid(AStatus, ChromMatchedAt) = 0) Then
Row34:	Grid((ChromSize + 2 + Grid(INSTANCECOUNT,
Row35:	ChromMatchedAt) + 1), ChromMatchedAt) = i
Row36:	End If
Row37:	If (MatchDClass(i, ChromMatchedAt) = False) Then
Row38:	Grid(AStatus, ChromMatchedAt) = 1
Row39:	End If
Row40:	Else
Row41:	GridRCounter = Insert(GridRCounter, chrom, i)
Row42:	End If
Row43:	Next
Row44:	Dim j As Integer
Row45:	ReDim t(1 To rub + 1)
Row46:	t(1) = 0
Row47:	For i = 1 To GridRCounter

Row48:	If ((Grid(AStatus, i) = 0) And ((Grid(DECISIONCLASS, i) = 1))) Then
Row49:	For j = 1 To Grid(INSTANCECOUNT, i)
Row50:	t(1) = t(1) + 1
Row51:	t(t(1) + 1) = Grid((ChromSize + 3 + j), i)
Row52:	Next
Row53:	End If
Row54:	Next
Row55:	CalculateUAI = 0
Row56:	End Function

Function takes input two parameters, i.e. Chrom array and XConcept value (in xc variable). In Row10, function declares a Grid array and sets initial Grid Record Count (GridRCounter) to zero. ChromSize represents the total number of attributes in Chrom array. DECISIONCLASS, INSTANCECOUNT and AStatus store the index of the column that stores decision class, index of column that stores the total number of objects in current decision class and finally the status of the object (either it has been considered or not). Note that this time, function offers a bit of optimization in a way that it first determines the total number of objects that belong to XConcept and declares Grid according to that.

In Row27, first record is inserted into the Grid. Values of all the attributes are inserted, instance count is set to '1', attribute status is set to zero, and index of the object (from dataset). The function then traverses all the records in the dataset from record number '2' to last record and for each record, it matches the value of the attributes in the dataset with those stored in grid. If values of the attributes are matched then the value of decision class is also checked. If the decision class also matches then the objects is stored in the next column in the same row (in grid) and instance count is incremented. Otherwise, if decision class is not matched then 'AStatus' attribute is set to '1' which means that all the objects with the same value of attributes will not be part of lower approximation. However, if the object with the same value of attributes is not matched with those stored in Grid, the object is inserted in the same way as first record and GridRCounter is incremented.

After traversing all the records in the dataset, the grid now contains values of conditional attributes (in row), the number of objects having these values, decision class and finally the indexes of all the objects that contain this decision class. Now from Row47 to Row54, we run two nested loops. First loop iterates through rows and inner loops iterate through columns in each row. In each row, first loop checks if the 'AStatus' column is '0' (which means that objects in this row belong to lower approximation) and decision class is same as that mentioned in XConcept (which means that objects belong to lower approximation of same decision class. The objects are stored in a temporary array 't'. You can store it in any global array or alternatively, array can be passed to function as a reference and this function will fill the array.

10.7 Upper Approximation Using Conventional Method

Upper approximation is calculated by 'FindUAO' method. This method is called from 'CalculateUAObjects' which is same as 'CalculateLAObjects', however, in final steps instead of calculating objects in equivalence class structure (using conditional attributes) which are a subset of XConcept set, we find those equivalence classes which have non-empty interaction with XConcept set. This is what is implemented in FindUAO method (Listing 10.20).

Listing 10.20: FindUAO Method
Row1: Function FindUAO(ByRef TCC() As Integer, cr As Integer)
Row2: Dim X As Integer
Row3: Dim IsNonNegative As Boolean
Row4: IsNonNegative = False
Row5: Dim i As Integer
Row6: For i = 2 To TCC(cr, 1) + 1
Row7: If (data(TCC(cr, i), DAtt) = XConcept) Then
Row8: IsNonNegative = True
Row9: End If
Row10: Next
Row11: If IsNonNegative = True Then
Row12: For X = 2 To TCC(cr, 1) + 1
Row13: UAObjects(1) = UAObjects(1) + 1
Row14: UAObjects(UAObjects(1) + 1) = TCC(cr, X)
Row15: Next
Row16: End If
Row17: End Function

It receives two arguments, i.e. reference of 'TotalCClasses' array and row index of current equivalence class. From Row6 to Row10, it iterates through equivalence class and checks whether any of the objects in 'TotalCClasses' (for current equivalence class) has the same decision class value as mentioned in the concept. If this is the case then all the objects belonging to this equivalence class will be part of upper approximation and will be assigned to 'UAObjects' array.

10.8 Upper Approximation Using Redefined Preliminaries

CalculateUAI method calculates upper approximation using redefined preliminaries. The method is called from the Main method. Following is the listing of this method. Following is the listing of this method (Listing 10.21).

Listing 10.21: CalculateUAIMethod	
Row1:	Function CalculateUAI(ByRef chrom() As Integer, ByRef xc As Integer) As Single
Row2	Dim DF, UC As Integer
Row3:	Dim i As Integer
Row4:	Dim GRC As Integer
Row5:	Dim ChromMatched As Boolean
Row6:	Dim DClassMatched As Boolean
Row7:	Dim ChromMatchedAt As Integer
Row8:	Dim DClassMatchedAt As Integer
Row9:	Dim NObject As Integer
Row10:	ReDim Grid(1 To cub + 3, 1 To rub)
Row11:	GridRCounter = 0
Row12:	ChromMatched = False
Row13:	DClassMatched = False
Row14:	ChromSize = UBound(chrom) - LBound(chrom) + 1
Row15:	DECISIONCLASS = UBound(chrom) + 1
Row16:	INSTANCECOUNT = DECISIONCLASS + 1
Row17:	AStatus = INSTANCECOUNT + 1
Row18:	XConcept = xc
Row19:	TDO = 0
Row20:	For i = 1 To rub
Row21:	If (data(i, DAtt) = XConcept) Then
Row22:	TDO = TDO + 1
Row23:	End If
Row24:	Next
Row25:	ReDim Grid(1 To ChromSize + 3 + TDO, 1 To rub) ' column,row
Row26:	If (GridRCounter = 0) Then
Row27:	GridRCounter = Insert(GridRCounter, chrom, 1)
Row28:	End If
Row29:	For i = 2 To rub
Row30:	ChromMatchedAt = MatchChrom(i, chrom, ChromMatched, GridRCounter) If (ChromMatched = True) Then
Row31:	Grid(INSTANCECOUNT, ChromMatchedAt) =
Row32:	Grid(INSTANCECOUNT, ChromMatchedAt) + 1
Row33:	Grid((ChromSize + 2 + Grid(INSTANCECOUNT, ChromMatchedAt) + 1), ChromMatchedAt) = i
Row34:	If (MatchDClass(i, ChromMatchedAt) = False) Then
Row35:	Grid(AStatus, ChromMatchedAt) = 1
Row36:	End If
Row37:	Else
Row38:	GridRCounter = Insert(GridRCounter, chrom, i)
Row39:	End If
Row40:	Next
Row41:	Dim j As Integer
Row42:	ReDim t(1 To rub + 1)
Row43:	t(1) = 0

```
Row44:    For i = 1 To GridRCounter
Row45:    Dim isnonnegative As Boolean
Row46:    isnonnegative = False
Row47:    For j = 1 To Grid(INSTANCECOUNT, i)
Row48:    If (data(Grid((ChromSize + 3 + j), i), DAtt) = XConcept) Then
Row49:    isnonnegative = True
Row50:    End If
Row51:    Next
Row52:    If isnonnegative = True Then
Row53:    For j = 1 To Grid(INSTANCECOUNT, i)
Row54:    t(1) = t(1) + 1
Row55:    t(t(1) + 1) = Grid((ChromSize + 3 + j), i)
Row56:    Next
Row57:    End If
Row58:    Next
Row59:    CalculateUAI = 0
Row60:    End Function
```

Just like 'CalculateLAI', 'CalulateUAI' takes the same arguments, i.e. reference of the chrom array and XConcept. Function performs the same steps and stores objects in the Grid with the same structure. However, note that now we have to calculate upper approximation, so all the objects having the same value of decision attributes and at least one of them leads to same decision class specified by XConcept will be part of upper approximation. So, note that we have skipped the condition to Check 'AStatus' only decision class is checked. But there is a complexity, an object inserted later (in grid) with same value of conditional attributes may lead to a different decision class (and update the decision class column as well) so, we have used two nested loops, first loop checks the decision class of each object (in Grid) to ensure that either of them leads to same decision class (as mentioned in XConcept), if any of such object is found, then all the objects in the same row may will be part of upper approximation.

10.9   Quick Reduct Algorithm

QuickReduct algorithm is one of the most famous feature selection algorithms, so we have provided the complete source code. Clicking the execute button will invoke QuickReduct method. Following is the listing of this method (Listing 10.22).

Listing 10.22: QuickReduct Method
Row1: Function QuickReduct () As Integer
Row2   GridRCounter = 0
Row3: Dim i, j As Integer
Row4: RUB = 6598
Row5: CUB = 168
Row6: ReDim data(1 To RUB, 1 To CUB)
Row7: For i = 1 To RUB
Row8:   For j = 1 To CUB
Row9:   data(i, j) = Cells(2 + i, 1 + j).Value
Row10:   Next j
Row11: Next i
Row12: OID = LBound(data, 2)
Row13: DAtt = UBound(data, 2)
Row14: Dim chrom() As Integer
Row15: ReDim chrom(1 To CUB - 2)
Row16: For i = 2 To CUB - 1
Row17: chrom(i - 1) = i
Row18: Next
Row19: Dim R() As Integer
Row20: Dim T() As Integer
Row21: Dim X() As Integer
Row22: Dim tmp(1 To 1) As Integer
Row23: Dim dp, sc, dp1, dp2, dp3 As Integer
Row24: ReDim TotalCClasses(1 To RUB, 1 To RUB + 1)
Row25: Call ClrTCC
Row26: Call setdclasses
Row27: dp = calculateDRR(chrom)
Row28: i = 1
Row29: Do
Row30: Range("A2").Value = i
Row31: Call Restore(T, R)
Row32: Call C_R(X, chrom, R)
Row33: For sc = 1 To UBound(X)
Row34: tmp(1) = X(sc)
Row35: Call ClrTCC
Row36: dp1 = calculateDRR(CUD(R, tmp))
Row37: If (dp1 > dp2) Then
Row38: T = CUD(R, tmp)
Row39: dp2 = dp1
Row40: End If
Row41: Range("A3").Value = sc
Row42: Range("A4").Value = dp1
Row43: Next
Row44: Call Restore(R, T)
Row45: Range("A5").Value = dp2
Row46:   i = i + 1
Row47: Loop Until (dp2 = dp)
Row48: Main = 1
Row49: End Function

Function loads the data in the same way as discussed previously, chrom is
initialized with attribute indexes. Note that we have taken TotalCClasses and
SetDClasses into QuickReduct method instead of 'CalculateDRR'. This measure is
taken to enhance the performance as a dependency will be calculated again and
again, so this will avoid the declaration of array again and again. However

'ClrTCC' will be called every time dependency will be calculated to clear the previously created equivalence class structure.

In Row27, we have calculated dependency using the entire set of conditional attributes. Row31 copies the array 'R' into 'T'. 'R' represents Reducts obtained so far. Row32 copies the attributes {Chrom − R} into {X} i.e. all the attributes of Chrom that are not in R will be copied to X. the loop at Row33 iterates through all the attributes in 'X'. Row35 clears the previous equivalence class structure. Row36 first combines the current attribute represented by 'tmp' array with 'R' and then passes this to calculateDRR function which actually calculates dependency that is stored in 'dp1'. Now if dp1 is greater than dp2 which represents the previous highest dependency (it means that combining current attribute with R has increased degree of dependency) then the current attribute is combined (concatenated) with R. dp2 is then updated by dp1. After the loop completes, we have 'R' combined with attribute that provides highest degree of dependency increase. So we copy 'T' (which contains highest degree attribute concatenated with R) into R. the process continues until the dp2 becomes equal to 'dp' the dependency of 'D' on entire set of conditional attributes.

10.9.1   Miscellaneous Methods

'CUD' method combines (concatenates) two integer arrays. It performs 'Union' operation. Following is the listing of the method (Listing 10.23).

Listing 10.23: CUD Method

```
Row1:    Function CUD(C() As Integer, D() As Integer) As Integer()
Row2     Dim csize As Integer
Row3:    Dim dsize As Integer
Row4:    Dim cd() As Integer
Row5:    If (isArrayEmpty(C)) Then
Row6:        Call Restore(cd, D)
Row7:        CUD = cd
Row8:        Exit Function
Row9:    End If
Row10:   If (isArrayEmpty(D)) Then
Row11:       Call Restore(cd, C)
Row12:       CUD = cd
Row13:       Exit Function
Row14:   End If
Row15:   ReDim cd(1 To UBound(C) + UBound(D))
Row16:   csize = UBound(C) - LBound(C) + 1
Row17:   dsize = UBound(D) - LBound(D) + 1
Row18:   Dim X As Integer
Row19:   For X = LBound(C) To UBound(C)
Row20:   cd(X) = C(X)
Row21:   Next
Row22:   Dim j As Integer
Row23:   X = X - 1
Row24:   For j = 1 To UBound(D)
Row25:       cd(X + j) = D(j)
Row26:   Next
Row27:   CUD = cd
Row28:   End Function
```

Function takes two arguments, i.e. two integer arrays and concatenates them. First it checks if array 'C' is empty then copies array 'D' into 'cd'. If 'D' is empty then 'C' is copied into 'cd'. If none of these cases exist, then we define 'cd' to be equal the size of 'C' plus size of 'D' and then copies both arrays in 'cd'.

10.9.2   Restore Method

Restores source array 'S' into target array 'T'. Following is the listing of the function (Listing 10.24)

Listing 10.24: Restore Method
Row1:   Function Restore(ByRef T() As Integer, ByRef S() As Integer) As Integer
Row2   If (isArrayEmpty(S)) Then
Row3:   Restore = 0
Row4:   Exit Function
Row5:   End If
Row6:   ReDim T(1 To UBound(S))
Row7:   Dim i As Integer
Row8:   i = 1
Row9:   While (i <= UBound(S))
Row10:   T(i) = S(i)
Row11:   i = i + 1
Row12:   Wend
Row13:   Restore = 1
Row14:   End Function

Function first checks whether the source array is empty in which case this function returns zero. Otherwise, it defines the target array to be equal to the size of source array 'S' and copies it to target array element by element.

10.9.3   C_R Method

This method copies one array into others by skipping the mentioned elements. Following is the listing of this function (Listing 10.25).

Listing 10.25: C_RMethod	
Row1:	Public Function C_R(tmp() As Integer, C() As Integer, R() As Integer) As Integer
Row2:	If (isArrayEmpty(R)) Then
Row3:	Call Restore(tmp, C)
Row4:	C_R = 0
Row5:	Exit Function
Row6:	End If
Row7:	ReDim tmp(1 To UBound(C) - UBound(R))
Row8:	Dim Tmpi, Ci, Ri As Integer
Row9:	Dim found As Boolean
Row10:	Tmpi = 1
Row11:	For Ci = 1 To UBound(C)
Row12:	found = False
Row13:	For Ri = 1 To UBound(R)
Row14:	If (C(Ci) = R(Ri)) Then
Row15:	found = True
Row16:	End If
Row17:	Next
Row18:	If (found = False) Then
Row19:	tmp(Tmpi) = C(Ci)
Row20:	Tmpi = Tmpi + 1
Row21:	End If
Row22:	Next
Row23:	C_R = 1
Row24:	End Function

It takes three arguments, 'tmp', the reference of the array in which final result will be copied, 'C' array which will be copied and 'R' array whose elements (also present in 'C') will be skipped.

If 'R' is empty, then entire of 'C' will be copied to 'tmp' using 'Restore' function, otherwise tmp is defined equal to size of 'C' minus size of 'R'. The function then iterates through each element in 'C', outer loop controls this. The currently selected element is then searched in 'R' (by the inner loop), if element is found, it is skipped otherwise it is copied to 'tmp'.

10.10   Summary

In this chapter we have presented source code of the provided API library and its description. All the API functions are explained up to sufficient details to use them or modify with any of the feature selection or RST based algorithm. Each function was explained with details of the tasks it provides and explanation of important statements and data structures used. For some complex data structure, diagrammatic description was also provided.

Chapter 11
Dominance Based Rough Set APIs Library

In this chapter we will present VBA source code for calculating approximations. We will calculate both lower and upper approximations alongwith dominance relation and class unions. The main intention of the chapter is to clear the programming logic behind calculating these measures. At this point it is referred to have some basic tutorial about VBA. The ready-to-take or ready-to-run code is given in 'DRSA_PL.bas' and 'DRSA_PU.bas' files where first one contains source of lower approximations and second one contains source code of upper approximations.

11.1 Lower Approximations

We will start with calculating $\underline{P}(Cl_t^{\leq})$ then we will move to $\underline{P}(Cl_t^{\geq})$. We will follow all three steps discussed in Sect. 9.2. So, first calculate $\underline{P}(Cl_t^{\leq})$. Following is listing of the 'Main' function (Listing 10.1).

© Springer Nature Singapore Pte Ltd. 2019
M. S. Raza and U. Qamar, *Understanding and Using Rough Set Based Feature Selection: Concepts, Techniques and Applications*,
https://doi.org/10.1007/978-981-32-9166-9_11

Listing 10.1: Main function
Row1: Function Main() As Integer
Row2: Dim i, j As Integer
Row3: RUB = 3196
Row4: CUB = 38
Row5: ReDim Data(1 To RUB, 1 To CUB)
Row6: For i = 1 To RUB
Row7: For j = 1 To CUB
Row8: Data(i, j) = Cells(2 + i, 1 + j).Value
Row9: Next j
Row10: Next i
Row11: OID = LBound(Data, 2)
Row12: DAtt = UBound(Data, 2)
Row13: Find_PL_L_t (0)
Row14: Dim str As String
Row15: str = ""
Row16: For j = 2 To pl(1)
Row17: If (pl(j) <> 0) Then
Row18: str = str + ",X" & pl(j)
Row19: End If
Row20: Next
Row21: Cells(2, 1).Value = str
Row22: Main = 1

Row2 to Row5 provide some data declarations. 'RUB' stands for 'Row Upper Bound' where 'CUB' stands for 'Column Upper Bound'. Both variables are intented to store maximum number of rows and columns in dataset. In our case there were '3196' rows and '38' columns. Row6 to Row9 load our data in 'Data' array. 'OID' and 'DAtt' represent ObjectID and Decision Attribute. These two variables store index of first column (in each dataset a column is added at start that contains serial number of each object. This serial number is used as ObjectID) and the attribute representing the decision class (probably the last attribute).

Row13 calls function 'Find_PL_L_t ()' with zero (0) as argument. This function calculates $\underline{P}(Cl_t^{\leq})$. Row14 to Row21 simply display the objects found in Cell (2,1). Final function quits at Row22.

11.1.1   Function: Find_PL_L_t ()

The listing of the function is given below (Listing 10.2)

Listing 10.2: Find_PL_L_t
Row1: Function Find_PL_L_t(t As Integer)
Row2 Call Get_Cl_LE_t(t)
Row3: Call DP_N_X
Row4: Find_P_L_G_T
Row5: End Function

This function takes 't' i.e. the index of the class for which we are going to calculate approximation. It works exactly according to the three steps mentioned in

Chap. 9. First it calculates $Cl_t^\leq$, then $D_P^-(x)$ and finally the objects belonging to lower approximation are actually calculated.

11.1.2 Function: Get_Cl_LE_t

Following code shows the listing of function 'Get_Cl_LE_t' (Listing 10.3).

```
Listing 10.3: Get_Cl_LE_t
Row1:    Function Get_Cl_LE_t(t As Integer)
Row2     ReDim cl(1 To rub * 2)
Row3:    Dim j As Integer
Row4:    j = 2
Row5:    Dim i As Integer
Row6:    For i = 1 To rub
Row7:      If (data(i, DAtt) <= t) Then
Row8:        cl(1) = j
Row9:        cl(j) = data(i, OID)
Row10:       j = j + 1
Row11:     End If
Row12:   Next
Row13:   End Function
```

This function calculates $Cl_t^\leq$. Row7 to Row11 form core of the function. Entire dataset is scanned for decision classes and for the objects having less preferred decision classes than 't', their ObjectIDs are collected in array named 'cl'. Note that first index of 'cl' contains total number of objects present in it.

11.1.3 Function: DP_N_X

Listing 10.4 below shows the source code of $D_P^-(x)$.

```
Listing 10.4: DP_N_X
Row1:    Function DP_N_X()
Row2     ReDim dp(1 To cl(1) - 1, 1 To rub + 1)
Row3:    Dim i As Integer
Row4:    Dim j As Integer
Row5:    Dim k As Integer
Row6:    Dim s As Boolean
Row7:    For i = 2 To cl(1)
Row8:      dp(i - 1, 1) = 2
Row9:      dp(i - 1, 2) = cl(i)
Row10:     For j = 1 To rub
Row11:       If (j <> cl(i)) Then
Row12:         s = compareL(j, cl(i))
Row13:         If (s <> True) Then
Row14:         dp(i - 1, dp(i - 1, 1) + 1) = j
Row15:         dp(i - 1, 1) = dp(i - 1, 1) + 1
Row16:       End If
Row17:     End If
Row18:   Next
Row19:   Next
Row20:   End Function
```

The function compares attributes of each object (present in 'cl' array) with other objects and all the objects dominated by current object are collected in 'dp' array. 'compareL()' function at Row12 actually compares the objects.

11.1.4 Function: Find_P_L_G_T

This function actually calculates the objects belonging to lower approximation. Source code of the function is given in Listing 10.5 below.

Listing 10.5: Find_P_L_G_T function
Row1:   Function Find_P_L_G_T()
Row2   Dim i As Integer
Row3:   Dim ie As Boolean
Row4:   Dim k As Integer
Row5:   Dim L As Integer
Row6:   Dim j As Integer
Row7:   ReDim pl(1 To cl(1))
Row8:   Dim plc As Integer
Row9:   pl(1) = 1
Row10:   For i = 1 To UBound(dp, 1)
Row11:   plc = 0
Row12:   If ((dp(i, 1) - 1) <= cl(1) - 1) Then
Row13:   For j = 2 To dp(i, 1)
Row14:   For k = 2 To cl(1)
Row15:   If (dp(i, j) = cl(k)) Then
Row16:   plc = plc + 1
Row17:   End If
Row18:   Next
Row19:   Next
Row20:   If (plc = dp(i, 1) - 1) Then
Row21:   For L = 2 To dp(i, 1)
Row22:   If (found(dp(i, L)) = False) Then
Row23:   pl(1) = pl(1) + 1
Row24:   pl(pl(1)) = dp(i, L)
Row25:   End If
Row26:   Next
Row27:   End If
Row28:   L = 0
Row29:   End If
Row30:   Next
Row31:   End Function

Each row of 'dp' contains $D_P^-(x)$ for each object in 'cl'. Now this function calculates $D_P^-(x) \subseteq Cl_t^{\leq}$. For this purpose all the objects in a single row of 'dp' are validated to be subset of $Cl_t^{\leq}$ and belonging objects are stored in 'pl'.

Now we will provide the source code for $\underline{P}(Cl_t^{\geq})$. We will only provide listing as the code works on the same pattern as explained in case of $\underline{P}(Cl_t^{\leq})$. Listing 10.6–10.8 show source code of these functions.

Listing 10.6: Find_PL_G_t function

Row1:	Function Find_PL_G_t(t As Integer)
Row2	Call Get_Cl_GE_t(t)
Row3:	Call DP_P_X
Row4:	Find_P_L_G_T
Row5:	End Function

Listing 10.7: Get_Cl_GE_t function

Row1:	Function Get_Cl_GE_t(t As Integer)
Row2	ReDim cl(1 To rub)
Row3:	Dim j As Integer
Row4:	j = 2
Row5:	Dim i As Integer
Row6:	For i = 1 To rub
Row7:	If (data(i, DAtt) >= t) Then
Row8:	cl(1) = j 'first index of cl contains the index of last element in array
Row9:	cl(j) = data(i, OID)
Row10:	j = j + 1
Row11:	End If
Row12:	Next
Row13:	End Function

Listing 10.8: DP_P_X function

Row1:	Function DP_P_X()
Row2	ReDim dp(1 To cl(1) - 1, 1 To rub + 1)
Row3:	Dim i As Integer
Row4:	Dim j As Integer
Row5:	Dim k As Integer
Row6:	Dim s As Boolean
Row7:	For i = 2 To cl(1)
Row8:	dp(i - 1, 1) = 2
Row9:	dp(i - 1, 2) = cl(i)
Row10:	For j = 1 To rub
Row11:	If (j <> cl(i)) Then
Row12:	s = compareG(j, cl(i))
Row13:	If (s <> True) Then
Row14:	dp(i - 1, dp(i - 1, 1) + 1) = j
Row15:	dp(i - 1, 1) = dp(i - 1, 1) + 1
Row16:	End If
Row17:	End If
Row18:	Next
Row19:	Next
Row20:	End Function

Note that the function Find_P_L_G_T() is the same as shown in previous listing. In 'Main' function (Listing 10.1) Row-13 will call the function 'Find_PL_G_t' and execution will start for calculation of $\underline{P}(Cl_t^{\geq})$.

11.2   Upper Approximations

The complete source code for calculating upper approximation is given in the 'DRSA_pu.bas' file. Source code is developed on the same basis as shown in case of calculating lower approximations along with same variable names in order to make things more understandable. All the three steps mentioned in Chap. 9 for calculating upper approximations are followed. Here we will be provide source listing of each of the function. First we will provide listings for $\bar{P}(Cl_t^{\leq})$ and then for $\bar{P}(Cl_t^{\geq})$.

Starting with 'Main' function (Listing 10.9), please note that this function here calls 'Find_PU_L_t'.

Listing 10.9: Main function	
Row1:	Function Main() As Integer
Row2	Dim i, j As Integer
Row3:	RUB = 3196
Row4:	CUB = 38
Row5:	ReDim Data(1 To RUB, 1 To CUB)
Row6:	For i = 1 To RUB
Row7:	For j = 1 To CUB
Row8:	Data(i, j) = Cells(2 + i, 1 + j).Value
Row9:	Next j
Row10:	Next i
Row11:	OID = LBound(Data, 2)
Row12:	DAtt = UBound(Data, 2)
Row13:	Find_PU_L_t (0)
Row14:	Dim str As String
Row15:	str = ""
Row16:	For j = 2 To pl(1)
Row17:	If (pl(j) <> 0) Then
Row18:	str = str + ",X" & pl(j)
Row19:	End If
Row20:	Next
Row21:	Cells(2, 1).Value = str
Row22:	Main = 1
Row23:	End Function

Listing 10.10 below shows the code of Find_PU_L_T function.

Listing 10.10: Find_PU_L_t function	
Row1:	Function Find_PU_L_t(t As Integer)
Row2	Call Get_Cl_LE_t(t)
Row3:	Call DP_N_X
Row4:	Find_P_L_G_T
Row5:	End Function

We have already provided description of 'Get_Cl_LE_t(t)' and 'DP_N_X()' functions. The logic and listing remains the same. In case of $\bar{P}(Cl_t^{\geq})$, the 'Main' function calls 'Find_PU_G_t' for which Listing 10.11 is given below.

Listing 10.11: Find_PU_G_t function	
Row1:	Function Find_PU_G_t(t As Integer)
Row2:	Call Get_Cl_GE_t(t)
Row3:	Call DP_P_X
Row4:	Find_P_L_G_T
Row5:	End Function

The only function that needs to be explained is 'Find_P_L_G_T'. The listing for this function is given below (Listing 10.12).

Listing 10.12: Find_P_L_G_T function	
Row1:	Function Find_P_L_G_T()
Row2	Dim i As Integer
Row3:	Dim ie As Boolean
Row4:	Dim k As Integer
Row5:	Dim L As Integer
Row6:	Dim j As Integer
Row7:	ReDim pl(1 To rub + 1)
Row8:	Dim plc As Integer
Row9:	Dim f As Boolean
Row10:	pl(1) = 1
Row11:	For i = 1 To UBound(dp, 1)
Row12:	plc = 0
Row13:	f = False
Row14:	For j = 2 To dp(i, 1)
Row15:	For k = 2 To cl(1)
Row16:	If (dp(i, j) = cl(k)) Then
Row17:	f = True
Row18:	Exit For
Row19:	End If
Row20:	Next
Row21:	If f = True Then
Row22:	Exit For
Row23:	End If
Row24:	Next
Row25:	If (f = True) Then
Row26:	For L = 2 To dp(i, 1)
Row27:	If (found(dp(i, L)) = False) Then
Row28:	pl(1) = pl(1) + 1
Row29:	pl(pl(1)) = dp(i, L)
Row30:	End If
Row31:	Next
Row32:	End If
Row33:	L = 0
Row34:	Next

The function calculates $D_P^-(x) \cap Cl_t^{\geq} \neq \emptyset$ and $D_P^+(x) \cap Cl_t^{\leq} \neq \emptyset$ depending on either you are calculating $\bar{P}(Cl_t^{\leq})$ or $\bar{P}(Cl_t^{\geq})$.

'cl' contains $Cl_t^{\geq}$ or $Cl_t^{\leq}$. 'dp' contains $D_P^-(x)$ or $D_P^+(x)$. Row16 actually checks if any object in 'dp' is equal to anyone in 'cl' to verify that either $D_P^-(x) \cap Cl_t^{\geq} \neq \emptyset$ or otherwise. If object is found i.e. the intersection of

$D_P^-(x)$ *and* $Cl_t^{\geq}$ is non empty, the loop terminates. Row25 to Row30 then copy the objects to array 'pl' which actually stores objects belonging to upper approximations.

11.3 Summary

In this chapter we have provided source code of the functions used for calculating lower approximations. Upper approximation is calculated on the same basis. It should be noted that during calculation of lower and upper approximations, dominance relation is also calculated and source code of dominance relation is given in this chapter as well. The intention was to clear the logic behind core preliminaries of DRSA so that it could be used on as-it-is basis or could be translated to some other language.

Printed in the United States
By Bookmasters